An Introduction to the Numerical Solution of Differential Equations

REVISED EDITION

ELECTRONIC & ELECTRICAL ENGINEERING RESEARCH STUDIES

APPLIED AND ENGINEERING MATHEMATICS SERIES

Series Editor: **Professor P. C. Kendall,** *Department of Electronic and Electrical Engineering, The University of Sheffield, England*

1. Convexity Methods in Variational Calculus
   Peter Smith

2. Spherical Harmonics and Tensors for Classical Field Theory
   M. N. Jones

3. An Introduction to the Numerical Solution of Differential Equations, Revised Edition
   Douglas Quinney

4. An Introduction to Fast Fourier Transform Methods for Partial Differential Equations, with Applications
   Morgan Pickering

An Introduction to the

Numerical Solution of Differential Equations

REVISED EDITION

Douglas Quinney
University of Keele, England

RESEARCH STUDIES PRESS LTD.
Letchworth, Hertfordshire, England
JOHN WILEY & SONS INC.
New York · Chichester · Toronto · Brisbane · Singapore

RESEARCH STUDIES PRESS LTD.
58B Station Road, Letchworth, Herts. SG6 3BE, England

Marketing and Distribution:

Australia, New Zealand, South-east Asia:
Jacaranda-Wiley Ltd., Jacaranda Press
JOHN WILEY & SONS INC.
GPO Box 859, Brisbane, Queensland 4001, Australia

Canada:
JOHN WILEY & SONS CANADA LIMITED
22 Worcester Road, Rexdale, Ontario, Canada

Europe, Africa:
JOHN WILEY & SONS LIMITED
Baffins Lane, Chichester, West Sussex, England

North and South America and the rest of the world:
JOHN WILEY & SONS INC.
605 Third Avenue, New York, NY 10158, USA

Library of Congress Cataloging-in-Publication Data

Quinney, Douglas, 1949–
An introduction to the numerical solution of differential equations.

(Electronic and electrical engineering research studies. Applied and engineering mathematics series; 3)
Spine title: The numerical solution of differential equations.
Includes index.
Bibliography: p.
1. Differential equations—Numerical solutions—Data processing. I. Title. II. Title: Numerical solution of differential equations. III. Series.
QA371.Q56 1987 515.3′5 87-4922
ISBN 0 471 91599 8 (Wiley)

British Library Cataloguing in Publication Data

Quinney, Douglas
An introduction to the numerical solution of differential equations.—Rev. ed.—
(Electronic & electrical engineering research studies. Applied and engineering mathematics series; 3).
1. Differential equations—Numerical solutions
I. Title. II. Series
515.3′5 QA371

ISBN 0 86380 055 6

ISBN 0 86380 055 6 (Research Studies Press Ltd.)
ISBN 0 471 91599 8 (John Wiley & Sons Inc.)

Printed in Great Britain by Galliard (Printers) Ltd., Great Yarmouth

Preface

The development of the electronic computer over the past three decades has brought about a revolution in the type and variety of numerical methods which can be applied to solve mathematical problems; the development of micro-computers has brought about a revolution within a revolution. The study of the methods which can be used on computing machines and the results they produce is called Numerical Analysis. This book is concerned with the application of numerical methods to produce approximate solutions for differential equations. It is intended to be used by research workers in science and engineering who require more than the brief details given in the documentation associated with many computer packages and may also prove useful as a basis for an advanced first course in the numerical solution of differential equations. The material arose from the author's practical research interests and also from courses given at the University of Keele. The approach is intended to be fairly informal, proofs of theorems are only given where to do so illustrates the application of the material presented. To this end there are approximately 100 worked examples and 120 exercises which the reader might wish to attempt. Where proofs are not given a reference to a more complete analysis is supplied.

Many mathematical models which attempt to interpret physical problems can be formulated in terms of the rate of change of one or more variables and as such naturally lead to differential equations. Furthermore, in addition to the differential equation the solution of the problem may have to satisfy one or more additional conditions and the way in which such conditions are specified greatly influences the type of method which can be applied. Most differential equations cannot be integrated to give a solution in terms of elementary functions and it may be necessary to integrate them numerically.

The great diversity of differential equations indicates that a variety of different methods may be needed, although a frequent approach is to replace the differential equation by a discrete equation. To give a basis for this type of method the first chapter discusses the fundamental role which is played by recurrence relations. Recurrence relations, also called difference equations, can be considered as a discrete analogue for differential equations. Beginning with some simple examples the solution of first and second order recurrence relations is discussed. Examples of the application of these methods to finding the zeros of a given function, to solving systems of linear equations, for determining eigenvalues and improving the accuracy of numerical quadrature are also examined.

Chapters 2 and 3 consider the solution of ordinary differential equations. The first of these chapters examines the solution of ordinary differential equations for which any supplementary conditions are imposed at a single point, i.e. initial value problems. The approach is based on single and multi-step difference methods. The consistency and convergence of these methods together with their stability when applied to a specific trial equation is considered. Chapter 3 extends the methods discussed for initial value problems to the solution of differential equations for which supplementary conditions are imposed at more than one point, i.e. boundary value problems. Two alternative methods are discussed. The first of these alternatives effectively replaces the differential equation by a system of discrete equations; these equations are then solved to determine an approximate solution. The second determines an approximate continuous solution which is defined throughout the interval where the solution is required. Methods for improving the computed solution are also examined.

The final 3 chapters discuss the solution of partial differential equations. Chapters 4 and 5 consider the solution of initial value problems in parabolic and hyperbolic equations and the connection between the methods employed and those used in Chapter 2 to solve ordinary differential equations of initial value type. Chapter 6 considers the solution of elliptic partial differential equations. Such equations are always of boundary value type and the methods introduced in this chapter may be likened to those in Chapter 3.

I would like to express my thanks to Peter Kendall for his invitation to write this book and for his encouragement and many helpful comments on previous drafts.

Keele, Staffordshire,
May, 1985.

Douglas Quinney.

Contents

Chapter 1. Recurrence Relations and Iterative Methods

Chapter 2. Ordinary Differential Equations Initial Value Problems

Chapter 3. Ordinary Differential Equations Boundary Value Problems

Chapter 4. Parabolic Partial Differential Equations

Chapter 5. Hyperbolic Partial Differential Equations

Chapter 5. Hyperbolic Partial Differential Equations (Ctnd)

Chapter 6. Elliptic Partial Differential Equations

CHAPTER 1
Recurrence Relations and Iterative Methods

Before studying the numerical methods which can be used to find approximate solutions for differential equations it is necessary to consider certain preliminaries. Amongst the many techniques for solving such problems some of the most useful are formulated in terms of recurrence relations, which are sometimes called difference equations and can be considered as discrete analogues of differential equations. We begin by examining very simple recurrence relations and demonstrate that even for simple problems it is necessary to proceed with care.

1.1 First Order Linear Recurrence Relations

Consider the problem of evaluating the integral

$$I_n = \int_0^1 x^n e^x \, dx, \qquad n \geqslant 0.$$

Integrating I_{n+1} by parts gives

$$I_{n+1} = [x^{n+1} e^x]_0^1 - \int_0^1 (n+1)x^n e^x \, dx$$

or

$$I_{n+1} = e - (n+1)I_n. \qquad (1.1.1)$$

Equation (1.1.1), often called a reduction formula, is an example of a simple recurrence relation. It can be used to generate the sequence of values I_n, n=1,2,..., given a suitable initial value I_0. As $I_0 = \int_0^1 e^x\,dx = e-1$, we have that $I_1=e-I_0=1$, $I_2=e-2I_1=e-2$ and so on.

In general, given a function G of m+1 arguments, an expression of the form

$$y_{n+1} = G(n, y_n, y_{n-1}, \ldots, y_{n-m+1}),\ n=m-1, m, \ldots, \qquad (1.1.2)$$

is called an mth order recurrence relation. If m initial values, y_0, y_1, ... ,y_{m-1}, are given then the recurrence relation (1.1.2) can be used to generate the sequence y_m, y_{m+1}, ... term by term. Any sequence of values, y_n, n=0,1,2,..., which satisfies (1.1.2) is called a solution of the recurrence relation. Furthermore, if y_n is a discrete function of n which involves m arbitrary constants and satisfies the recurrence relation (1.1.2) then it is called the general solution.

The general linear recurrence relation of the first order may be written in the form

$$y_{n+1} = a_n y_n + b_n. \qquad (1.1.3)$$

If $b_n=0$, for all n, the recurrence is said to be homogeneous in which case $y_1=a_0y_0$, $y_2=a_1a_0y_0$ and so on. In general the solution, y_n, is given by

$$y_n = a_{n-1}a_n \ldots a_0\, A,$$

where A is an arbitrary constant. If $b_n \neq 0$ then it can be shown that the general solution of (1.1.3) is obtained by adding any solution of the inhomogeneous equation to the general solution of the homogeneous equation. This is best demonstrated by an example.

Example 1.1.1 Find the general solution of $(n+2)y_{n+1}=(n+1)y_n+2n+1$.

Solution. (i) Consider the homogeneous recurrence relation $(n+2)y_{n+1}=(n+1)y_n$ from which

$$(n+1)y_n=ny_{n-1},\ ny_{n-1}=(n-1)y_{n-2},\ \ldots\ldots\ ,\ 2y_1=y_0,$$

and so

$$(n+1)y_n = y_0 \quad \text{or} \quad y_n = {}^{y_0}/_{(n+1)}.$$

As no initial condition is specified we may select y_0 as we wish, and so the general solution of $(n+2)y_{n+1}=(n+1)y_n$ is given by $y_n={}^{A}/_{(n+1)}$, where A is an arbitrary constant. That this is the general solution of the homogeneous equation can easily be checked by substituting this expression back into the equation and verifying that it does satisfy it for every value of A.

(ii) The next step is to find a solution of the inhomogeneous equation $(n+2)y_{n+1}=(n+1)y_n+2n+1$. As the inhomogeneous term is a linear function of n we try an expression of the form $y_n=an+b$. In order for this to be a solution it is necessary that

$$(n+2)(a(n+1) + b) = (n+1)(an + b) + 2n + 1$$

for all values of n, or equivalently,

$$2an + b + 2a = 2n + 1.$$

Hence, a=1 and b+2a=1 from which b=-1. The required particular solution is therefore $y_n=n-1$ which added to the general solution of the homogeneous equation gives the general solution of the inhomogeneous recurrence relation, namely,

$$y_n = {}^{A}/_{(n+1)} + n - 1.$$

It is worth the effort of substituting this expression into the original equation to show that this is the general solution.

When generating the solution of a first order recurrence relation term by term it is necessary to supply one member of the sequence $y_0, y_1, \dots$ in order to begin the recurrence; for example, with the reduction formula (1.1.1) it was necessary to specify I_0. Such an initial condition can be used to determine a value for the arbitrary constant in the general solution.

Example 1.1.2 Find the solution of the recurrence relation

$$(n+2)y_{n+1} = (n+1)y_n + 2n + 1$$

which satisfies $y_0=0$.

Solution. For n=0 the recurrence relation gives $2y_1 = y_0 + 1$, and so $y_1 = {}^1/_2$.

For n=1, $3y_2 = 2y_1+3 = 4$, or $y_2 = {}^4/_3$.
For n=2, $4y_3 = 3y_2+5 = 9$, or $y_3 = {}^9/_4$, and so on.

Clearly, it would be better if it were possible to express y_n directly in terms of n. The general solution of the recurrence relation is

$$y_n = {}^A/_{(n+1)} + n - 1,$$

from which $y_0=A - 1$, and so the initial condition, $y_0=0$, is satisfied provided A=1. Therefore, the required solution is

$$y_n = {}^1/_{(n+1)} + n - 1.$$

Exercise 1.1.1 Find the general solutions of the following first order recurrence relations.

(i) $y_{n+1} = ry_n$, r an arbitrary constant,
(ii) $y_{n+1} = ry_n + n$,
(iii) $ny_{n+1} = (1-n)y_n + n^2$.

In many cases, although it may not be possible to find an expression for the general solution, yet given y_0, it is quite easy to find the successive values y_1, y_2, ... by substituting into the recurrence relation. However, any error in y_n will influence the computed values for y_{n+1}, y_{n+2},.... and the growth of such errors needs careful examination.

Example 1.1.3 Compute a value for $I_{10} = \int_0^1 x^{10} e^x dx$ using the recurrence relation (1.1.1) with $I_0=e-1$.

Solution. Taking $I_0=1.71828183$ gives the results in column (a) of Table 1.1. In columns (b), (c) and (d) results are given for other "approximations" for e-1.

Table 1.1. Solutions of equation (1.1.1) for various initial values.

	(a)	(b)	(c)	(d)
I_0	1.71828183	1.71828000	1.70000000	1.80000000
I_1	1.00000000	1.00000180	1.00028180	0.91828183
I_2	0.71828183	0.71827817	0.68171817	0.88171817
:	:	:	:	:
I_5	0.39559929	0.39581890	2.58941890	-9.41058110
:	:	:	:	:
I_{10}	0.23566016	-6.40752211	-66340.8690	296539.127

This example demonstrates the danger in applying recurrence relations without sufficient prior analysis. A small change in the initial value for I_0 has produced a large change in the solution and so the problem is said to be **ill-conditioned**. Whether this ill-conditioning is inherent within the problem or induced by the method of solution needs to be determined. For this example we shall show that the ill-conditioning is induced by the method of solution.

Consider the general first order recurrence relation (1.1.3) together with a given initial value y_0. Now compute a second sequence z_0, z_1, z_2 satisfying the same recurrence relation subject to $z_0=y_0+\varepsilon$ and compare the values of y_n and z_n, i.e.

$$z_1 = a_0(y_0 + \varepsilon) + b_0 = a_0y_0+b_0+a_0\varepsilon = y_1 + a_0\varepsilon$$

$$z_2 = a_1(y_1+a_0\varepsilon) + b_1 = a_1y_1 + b_1 + a_1a_0\varepsilon$$

and in general

$$z_n = y_n + a_{n-1}a_{n-2} \cdots a_0\varepsilon.$$

Clearly, after n applications of the recurrence relation the original error ε has been amplified by a factor $a_{n-1}a_{n-2}\cdots a_0$. Therefore, if $|a_n|\leqslant 1$, for all n, the difference $|y_n - z_n|$ remains bounded and the recurrence relation is said to be **absolutely stable**; otherwise it is said to be **unstable.** In the previous example $a_n=n+1$ and so any error, ε, in I_0 produces an error $10.9.8.7.6.5.4.3.2.\varepsilon$ = $3{,}628{,}800\varepsilon$ in I_{10}; the recurrence relation is absolutely unstable. Even if I_0 were given exactly, which in this case is not possible since e is irrational, further errors will be introduced whenever the recurrence relation is used. If the solution y_n is increasing then even though the absolute error grows $|y_n-z_n|$ may be small in comparison with y_n; in which case it is the **relative error**, i.e. the ratio of the absolute error to the computed solution, which is of importance. If the relative error remains bounded then the recurrence relation is said to be **relatively stable.**

Thus far we have only considered the ill-conditioning of the first order recurrence relation (1.1.3) with respect to any additional initial condition which is supplied to obtain a specific solution. It is also possible to consider whether the problem is ill-conditioned with respect to the coefficients a_n and b_n; full details are given by Fox and Mayers (1968).

Exercise 1.1.2 Find the solution of $y_{n+1}=(1+\alpha)y_n+1$, which satisfies $y_0=1$. For which values of α is this recurrence relation (i) absolutely stable, (ii) relatively stable?

1.2 Second Order Linear Recurrence Relations

A simple example of a linear second order recurrence relation is given by the Fibonacci sequence. Let $y_0 = 1$ and $y_1 = 1$, then generate the sequence y_2, y_3, ... using the recurrence relation

$$y_{n+1} = y_n + y_{n-1}.$$

This gives the sequence of values $y_2=y_1+y_0=2$, $y_3=y_2+y_1=3$, $y_4=5$, $y_5=8$, $y_6=13$ and so on. This is an example of a linear second order recurrence relation. In its most general form a second order linear recurrence relation is given by

$$y_{n+1} + a_n y_n + b_n y_{n-1} = c_n. \qquad (1.2.1)$$

If $c_n=0$, for all n, such an equation is said to be homogeneous. If a_n and b_n are constants then it is relatively easy to find the general solution of the homogenous equation by looking for solutions of the form $y_n=Ar^n$, where A and r are constants. Substituting this expression into the recurrence relation gives

$$Ar^{n+1} + aAr^n + bAr^{n-1} = 0$$

where for the sake of clarity we have put $a_n=a$ and $b_n=b$. From this equation it is clear that A is arbitrary; but that r must satisfy the quadratic equation

$$r^2 + ar + b = 0. \qquad (1.2.2)$$

Equation (1.2.2) is called the **auxiliary equation** and has two solutions, r_1 and r_2 say. If $r_1 \neq r_2$ then the general solution of the homogeneous form of (1.2.1) is given by $y_n=Ar_1^n+Br_2^n$, where A and B are arbitrary constants; on the other hand, if $r_1=r_2$ then $y_n=(A+nB)r_1^n$ is the general solution.

If the coefficients a_n and b_n in equation (1.2.1) are functions of n there is no general technique for solving the problem.

Exercise 1.2.1 Find the general solution of

$$\text{(i)}\quad y_{n+1} + y_n - 6y_{n-1} = 0,$$
$$\text{(ii)}\quad y_{n+1} + 2y_n + y_{n-1} = 0.$$

The general solution of the recurrence relation (1.2.1) when $c_n=0$ is given by adding any solution of the inhomogeneous equation to the general solution of the homogeneous form. A solution can sometimes be found by inspection; for example, if c_n is a polynomial of degree r in n the required solution may be of the form

$$y_n = \alpha_r n^r + \alpha_{r-1} n^{r-1} + \ldots + \alpha_0$$

where $\alpha_r, \alpha_{r-1}, \ldots, \alpha_0$ can be found by substituting this expression back into the recurrence relation.

Example 1.2.1 Find the general solution of $10y_{n+1}-101y_n+10y_{n-1}=n$.

Solution. Firstly we find the general solution of the homogeneous form of the equation, i.e. $10y_{n+1}-101y_n+10y_n=0$. For $y_n=Ar^n$ to be a solution it is necessary that

$$10Ar^{n+1} - 101Ar^n + 10Ar^n = 0.$$

Therefore, A is arbitrary but r must satisfy the auxiliary equation

$$10r^2 - 101r + 10 = 0,$$

giving r=0.1 or r=10. Thus, the general solution is then given by

$$y_n = A10^n + B10^{-n},$$

where A and B are arbitrary constants. Next, it is necessary to find a solution of the inhomogeneous equation. Since the right-hand side is a linear function of n we try, as a possible solution, $y_n=\alpha_1 n+\alpha_0$. Substituting this expression into the inhomogeneous equation gives

$$10(\alpha_1(n+1)+\alpha_0)-101(\alpha_1 n+\alpha_0)+10(\alpha_1(n-1)+\alpha_0)=n$$

or

$$-81\alpha_1 n - 81\alpha_0 = n,$$

and comparing coefficients on the right and left of the equality gives $\alpha_1=-1/81$ and $\alpha_0=0$. Therefore, the general solution of the recurrence relation

$$10y_{n+1} - 101y_n + 10y_{n-1} = n$$

is

$$y_n = A10^n + B10^{-n} - n/81.$$

Exercise 1.2.2 Find the general solution of

(i) $y_{n+1} + y_n - 6y_{n-1} = 4n^2 - 14n,$

(ii) $y_{n+1} = y_n + y_{n-1}.$

The general solution of a second order recurrence relation will contain two arbitrary constants, A and B say, which can only be determined if two additional conditions are imposed on the sequence of values y_0, y_1,.. . If y_0 and y_1 are given the recurrence relation can be used to find y_2, then y_3 and so on. Since y_0 and y_1 are used to find the rest of the sequence this problem is said to be of **Initial Value Type.** If y_0 and y_N, N>1, are given then the problem is said to be of **Boundary Value Type.** In either case, any discrepancy in specifying such conditions precisely will produce corresponding errors in A and B. For example, with linear constant coefficient recurrence relations, if such errors are δA and δB, the error in y_n will be $\delta A r_1^n + \delta B r_2^n$ if $r_1 \neq r_2$ and $(\delta A + n\delta B)r_1^n$ otherwise. It is clear that in both cases if either $|r_1|$ or $|r_2|$ exceeds 1 then the error will grow. Therefore, second order recurrence relations for which either root of the auxiliary equation has modulus greater than one are absolutely unstable.

Example 1.2.2 Find the solution of $10y_{n+1}-101y_n+10y_{n-1}=n$ subject to (i) $y_0=0$, $y_1=1$, (ii) $y_0=y_{10}=0$.

Solution. (i) For n=1

$$10y_2 - 101y_1 + 10y_0 = 1, \text{ so } 10y_2 - 101 = 1 \text{ giving } y_2=10.2.$$

For n=2

$$10y_3 - 101y_2 + 10y_1 = 2, \text{ so } 10y_3 = 2 + 101\times 10.2 - 10 = 1022.2$$

giving $y_3 = 102.22$. Clearly, it would be more efficient to express y_n directly in terms of n. This can be achieved using the general solution obtained in Example 1.2.1, namely,

$$y_n = A\ 10^n + B\ 10^{-n} - {}^n/_{81}.$$

The initial conditions $y_0=0$ and $y_1=1$ give:-

for n=0, $y_0 = A + B - {}^0/_{81}$, hence $A + B = 0$,

for n=1, $y_1 = 10A + 0.1B - {}^1/_{81} = 1$, hence $100A + B = 10(1 + {}^1/_{81})$.

Therefore,

$$A = {}^{82\times 10}/_{81\times 99} \simeq 0.1022571....,$$
$$B = -A \simeq -0.1022571....,$$

and so

$$y_n = A\ (10^n - 10^{-n}) - n/81.$$

Suppose instead that A were taken to be 0.10226, i.e. correct to 5 decimal places, then the error in y_n would be approximately $2.8608\times 10^{-6}(10^n - 10^{-n})$ or about $2.8608\times 10^{n-6}$ if n is large. This error will grow rapidly demonstrating that the recurrence relation is absolutely unstable. Nevertheless, the solution itself will be much larger and the error will form only a small part of the computed value for y_n. Accordingly the problem is relatively stable.

(ii) Consider now the boundary conditions $y_0=y_{10}=0$. It would appear that it is not possible to use these conditions to find the solution term by term but using the general solution with n=0 and n=10 gives:-

for n=0, $y_0 = A + B - {}^0/_{81}$, hence $A + B = 0$,

for n=10, $y_{10}=10^{10}A + 0.1^{10}B - {}^{10}/_{81}$, hence $10^{20}A + B = 10^{11}/_{81}$.

Eliminating B produces

$$A = \frac{10^{11}}{81(10^{20} - 1)} \simeq 1.2345678\times10^{-11}.$$

If A and B were taken correct to 10 decimal places then they are effectively zero and substituting this value into the general solution gives the discrete function $y_n={}^{-n}/_{81}$. However, this function does not satisfy the boundary condition at n=10. This demonstrates the ill-conditioning from which this recurrence relation can suffer.

In practical situations where it may not be possible to find the general solution it may be necessary to apply the recurrence relation directly.

Example 1.2.3 Find the solution of $(n+1)y_{n+1}+2ny_n+(n-1)y_{n-1}=4n^2+2$, subject to (i) $y_0=10$, $y_1=-9$, (ii) $y_0=2$, $y_{10}=12$.

Solution. (i) This recurrence relation has coefficients which depend upon n. Suppose an expression of the form $y_n=Ar^n$, where A and r are constants, is substituted into the recurrence relation. Then in order for it to be a solution we must have

$$(n+1)r^2 + 2nr + (n-1) = 0.$$

This implies that r depends on n and contradicts the assumption that r is a constant; therefore, there can be no solution of this form.

As there is no general way of solving such a recurrence relation we apply the initial conditions directly to obtain the following results:-

n	0	1	2	3	4	5	6	7	8	9	10
y_n	10	-9	12	-7	14	-5	16	-3	18	-1	20

(ii) For the boundary conditions $y_0=2$, $y_{10}=12$ the above procedure is not applicable; however, the linearity of the equation can be exploited in another way. Consider the sequences of values generated by

$$\text{(a)}\quad (n+1)u_{n+1} + 2nu_n + (n-1)u_{n-1} = 4n^2+2,\ u_o=2, u_1=0,$$

and

$$\text{(b)}\quad (n+1)v_{n+1} + 2nv_n + (n-1)v_{n-1} = 0,\ v_o=0, v_1=1.$$

Now let $y_n=u_n + \alpha v_n$. Clearly y_n satisfies the original recurrence relation for all values of α; this is what linearity means. Furthermore,

$$y_o=u_o + \alpha v_o = u_o = 2,$$

and

$$y_{10}=u_{10} + \alpha v_{10} = 12 \quad \text{provided} \quad \alpha = \frac{(12 - u_{10})}{v_{10}}.$$

From (a) and (b) we obtain $u_{10}=11$, $v_{10}=-1$, which gives $\alpha=-1$ and so the required solution is given by $y_n=u_n-v_n$, $n=0,1,\ldots,10$.

n	0	1	2	3	4	5	6	7	8	9	10
y_n	2	-1	4	1	6	3	8	5	10	7	12

However, this method should be used with care as it may suffer from severe instability if the recurrence relations are unstable. Notice also, that if v_{10} were small then α may be subject to additional computational errors.

Exercise 1.2.3 Find the solution of $y_{n+1}+y_n-6y_{n-1}=4n^2-14n$ subject to (i) $y_0=0, y_1=0$, (ii) $y_1=0$, $y_5=1$, (iii) $y_0=y_{10}=0$.

Exercise 1.2.4 Find the solution of $y_{n+1}=0.01y_{n-1}$ subject to (i) $y_0=0, y_1=0.1$, (ii) $y_0=y_{10}=0$.

Exercise 1.2.5 Which, if any, of the recurrence relations in Exercises 1.2.3 and 1.2.4 are stable?

Exercise 1.2.6 Show that the ratio of successive terms in the Fibonacci sequence generated by $y_{n+1}=y_n+y_{n-1}$ subject to $y_0=y_1=1$ tends to $(1+\sqrt{5})/2$.

Exercise 1.2.7 Use the method of example 1.2.3 to find solutions of $y_{n+1} + y_n - 6y_{n-1} = 4n^2 - 14n$ subject to $y_0=y_{10}=0$. How do you account for the inaccuracy of the solution? (Hint: Consider the stability of the recurrence relation.)

1.3 Non-Linear Recurrence Relations for Solving f(x)=0

In this section we shall consider some simple examples of non-linear recurrence relations. Let f be any continuous function then a value of x for which f vanishes is called a **zero** of f or equivalently x is said to be a **root** or solution of the equation $f(x)=0$. For example, it might be necessary to find the solutions of

$$x^2 - 3x + 2 = 0$$

or

$$x^5 - 3x + 2 = 0$$

or possibly

$$e^{-x} - x = 0.$$

The first of these examples is relatively easy to solve because the roots can be found by inspection or by using the "quadratic formula", i.e. the roots of the general quadratic equation $ax^2+bx+c=0$ are given by

$$x = \frac{-b \pm \sqrt{(b^2 - 4ac)}}{2a}.$$

Unfortunately, even such simple problems are subject to induced errors.

Example 1.3.1 Find the solutions of $x^2+400x+0.01=0$.

Solution. Using a calculator and working to 8 significant figures the solutions are given by $x=(-400\pm399.99994)/2$ i.e. $x=-399.99997$ and $x=-3\times10^{-5}$. The product of the roots should give 0.01, but multiplying these two values produces 0.011999. This discrepancy is caused by subtracting two very nearly equal numbers to give the smallest root and produces a dramatic reduction in the accuracy of the solution. It is better to obtain the smaller solution as $0.01/(-399.99997)$ which gives the correct result -2.5×10^{-5}.

For many equations of the form $f(x)=0$ there is no closed form for the solution and so it may be necessary to resort to numerical techniques. As a simple example consider the equation

$$x^2 - 3x + 2 = 0$$

which may be written in an equivalent form as $x = (x^2 + 2)/3$. In the latter form it can be used to generate a sequence x_1, x_2, ... from the recurrence relation

$$x_{n+1} = (x_n^2 + 2)/3, \quad n=0,1,2, \ldots,$$

provided an initial value x_0 is supplied. For example, taking $x_0=0$ gives

$$x_1=0.666666667, \quad x_2=0.814814815,$$
$$x_3=0.887974394, \quad x_4=0.929499508,$$
$$x_5=0.954656440, \quad x_6=0.970456310,$$
$$x_7=0.980595150, \quad x_8=0.987188949,$$
$$x_9=0.99151400,$$

$$\vdots$$

$$x_{20}=0.999854663,$$

$$\vdots$$

$$x_{30}=0.999997481.$$

It would appear that this sequence is converging to 1, which is one of the solutions of $x^2-3x+2=0$. Obviously this is not the only possible rearrangement of the original equation. Another possible choice is $x^2 = 3x - 2$ giving the recurrence relation

$$x_{n+1} = \sqrt{(3x_n - 2)},$$

alternatively

$$x = (3x - 2)/x$$

gives

$$x_{n+1} = (3x_n - 2)/x_n = 3 - 2/x_n.$$

There are many other possibilities. We are therefore led to ask which method, if any, can be guaranteed to work. More to the point, which method will work the fastest in the sense of solving the problem and yet require the least amount of effort.

Exercise 1.3.1 Determine to which root of $x^2 - 3x + 2 = 0$ the following methods converge:-

(i) $x_{n+1} = \sqrt{(3x_n - 2)}$, $x_0=3$,
(ii) $x_{n+1} = 3 - {}^2/x_n$, $x_0=3$.

Each of the methods suggested for solving the equation $x^2-3x+2=0$ has taken the form $x_{n+1} = G(x_n)$. Such methods are called **iterative** and G is called the **iterative function**, but they are nothing more than simple non-linear recurrence relations. The quest now is to find conditions under which it is possible to guarantee convergence of an iterative process to a required solution. Firstly, notice that if the required solution is at $x=\mathbf{a}$ then $G(\mathbf{a})=\mathbf{a}$; that is a is a fixed point of G. Now write the difference between the n-th iterate, x_n, and a as e_n then

$$e_{n+1} = x_{n+1} - \mathbf{a} = G(x_n) - G(\mathbf{a}) = G(\mathbf{a}+e_n) - G(\mathbf{a}). \quad (1.3.1)$$

Using the remainder form of the Taylor series expansion of G about **a** gives

$$G(\mathbf{a}+e_n) = G(\mathbf{a}) + e_n G'(\xi),$$

where ξ lies between **a** and $\mathbf{a}+e_n$, and so

$$e_{n+1} = G'(\xi)e_n. \quad (1.3.2)$$

If there exists a constant C such that $|G'(x)| \leqslant C < 1$ near $x=\mathbf{a}$ then

$$|e_{n+1}| \leqslant C|e_n| \leqslant C^2|e_{n-1}| \leqslant \dots \leqslant C^{n+1} |e_0|,$$

and e_{n+1} tends to zero. In this case x_n tends to the solution **a** as required. As there is a linear relationship between e_n and e_{n+1} the iteration

$$x_{n+1} = G(x_n)$$

is said to **converge linearly** to a root $x=\mathbf{a}$ provided $|G'(x)|<1$ in some interval containing the solution **a** and an initial estimate x_0. (We have assumed that $G'(\mathbf{a}) \neq 0$.)

For the equation $x^2-3x+2=0$ and the recurrence relation

$$x_{n+1} = (x_n^2 + 2)/3$$

the iterative function is

$$G(x) = (x^2 + 2)/3$$

for which $G'(x) = 2x/3$. If $|x| < 3/2$ then $|G'|<1$ and so this iteration will converge linearly to $x=1$ from any x_0 in this interval. Near $x=2$, $|G'| \simeq 4/3$ which lies outside the interval $(-1,1)$ and the iteration does not converge to the second solution at $x=2$. For the same equation, and the re-arrangement

$$x_{n+1} = \sqrt{(3x_n - 2)}$$

we have

$$G(x) = \sqrt{(3x - 2)}$$

and

$$G'(x) = 3/[2\sqrt{(3x-2)}].$$

Near 2, $|G'| \simeq 0.75$ which satisfies the condition for linear convergence, but near x=1, $|G'| \simeq 1.5$ and the iteration does not converge. Finally, still for the same quadratic, the iteration

$$x_{n+1} = 3 - 2/x_n$$

has

$$G(x)=3 - 2/x,$$

and

$$G'(x) = 2/x^2.$$

Near x=2, $|G'| \simeq 1/2$ which satisfies the condition for linear convergence; however, near x=1, $|G'| \simeq 2$, and this iteration does not converge to this solution.

Example 1.3.2 Find the solution of $e^{-x} - x = 0$.

Solution. Sketching the curves $y = e^{-x}$ and $y = x$ it is clear that the required solution lies between x=0 and x=1. (See Figure 1.3.1.)

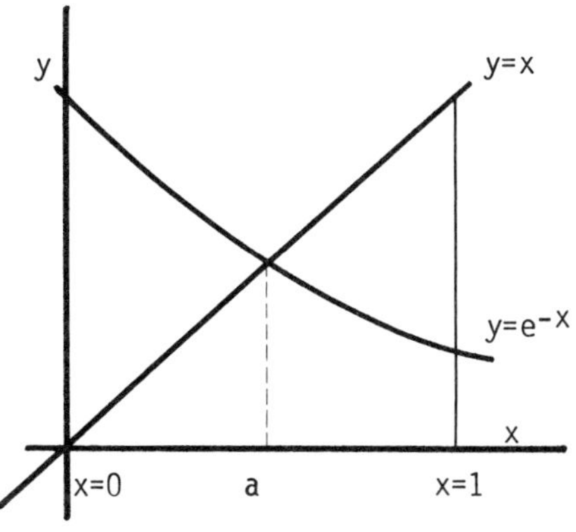

Figure 1.3.1

Now consider the iteration $x_{n+1} = \exp(-x_n)$ for which $G(x)= e^{-x}$ and $G'(x)=-e^{-x}$. Since $|G'(x)| < 1$ for all $x > 0$ this iteration will converge linearly to the solution from any $x_0>0$. Taking $x_0 = 1$ gives

$$x_1 = 0.3678794, \quad x_2 = 0.6922006,$$
$$x_3 = 0.5004735, \quad x_4 = 0.6062435,$$
$$\vdots$$
$$x_{19} = 0.567136, \quad x_{20} = 0.567148,$$

and so to 4 decimal places the required solution lies at $x=0.5671$.

Exercise 1.3.2 Find all three real solutions of $x^5-3x+2=0$ using suitable linearly convergent methods.

The previous examples show that a linearly convergent process for determining the solutions of $f(x)=0$ is relatively slow. For the equation $x^2-3x+2=0$ both

$$x_{n+1} = \sqrt{(3x_n - 2)}$$

and

$$x_{n+1} = 3 - 2/x_n$$

converge to the larger solution but the former has a value for $|G'|\simeq 0.75$ whilst the latter has a corresponding value of 0.5; clearly the latter method will be the faster. In general the smaller $|G'(\mathbf{a})|$ the more rapidly the process will converge; however, if the limiting value of $G'(\mathbf{a}) = 0$ is taken the previous analysis is no longer valid and it is necessary to look at higher order terms. From equation (1.3.1), expanding $G(\mathbf{a}+e_n)$ in a Taylor series about a and using $G(\mathbf{a})=\mathbf{a}$ we have that provided G is sufficiently smooth there exists a constant ξ such that

$$e_{n+1} = e_n G'(\mathbf{a}) + e_n^2 G''(\xi)/2,$$

which reduces to $e_{n+1}=e_n^2 G''(\xi)/2$ if $G'(\mathbf{a})=0$. If there exists a constant C such that $|G''(x)/2|\leq C$ near $x=\mathbf{a}$ then

$$|e_n| \leq C|e_{n-1}|^2 \leq C\ |C(e_{n-2})^2|^2 = C^3|e_{n-2}|^4$$
$$\ldots\ldots \leq (C|e_0|)^{2n}/C.$$

Therefore, if $C|e_0| < 1$ the error e_n tends to zero and the iteration $x_{n+1}=G(x_n)$ converges to **a**. Furthermore, since

$$e_n \simeq [G''(\xi)/2]\ (e_{n-1})^2; \qquad (1.3.3)$$

near x=**a**, such convergence is said to be **quadratic.**

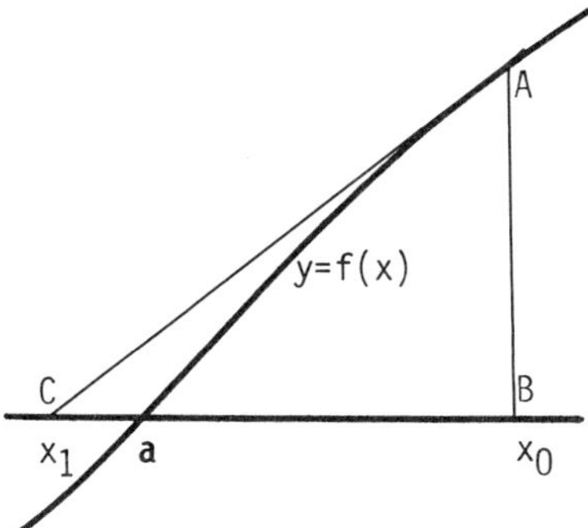

Figure 1.3.2

One quadratically convergent method can be obtained as follows. Consider a function f and an approximation x_0 for a zero at x=**a**. The slope of the curve y = f(x) at x_0 is given by

$$\tan ACB = \frac{AB}{BC} = \frac{f(x_0)}{x_0 - x_1},$$

where the x co-ordinate of the point C is x_1. However, from Figure 1.3.2, tan ACB = $f'(x_0)$, and rearranging this expression gives

$$x_1 = x_0 - f(x_0)/f'(x_0).$$

If this procedure is repeated so that a sequence of values x_1, x_2, x_3,... is generated by the recurrence relation

$$x_{n+1} = x_n - f(x_n)/f'(x_n), \qquad (1.3.4)$$

then x_n converges quadratically to a zero of f under quite simple conditions. This method is called the **Newton-Raphson** method or is more frequently referred to as **Newton's Method.**

The iterative function associated with Newton's method is given by

$$G(x) = x - {}^{f(x)}/_{f'(x)}.$$

Differentiating with respect to x gives

$$G'(\mathbf{a}) = 1 - {}^{[f'(\mathbf{a})f'(\mathbf{a}) - f(\mathbf{a})f''(\mathbf{a})]}/_{[f'(\mathbf{a})f'(\mathbf{a})]}.$$

Therefore, since $f(\mathbf{a})=0$, $G'(\mathbf{a}) = 0$ provided $f'(\mathbf{a})\neq 0$. The second derivative of G, near $x=\mathbf{a}$, is approximately $G''(\mathbf{a})={}^{f''(\mathbf{a})}/_{f'(\mathbf{a})}$, therefore, Newton's method will converge quadratically to a provided

$$\left|e_0\ {}^{f''(x)}/_{f'(x)}\right| < 2,$$

for all x sufficiently close to **a**.

Example 1.3.3 Find the solution of $e^{-x} = x$ using Newton's method.

Solution. Here $f(x)=e^{-x}-x$, $f'(x)=-e^{-x}-1$, $f''(x)=e^{-x}$ and hence $\left|{}^{f''(x)}/_{f'(x)}\right| < 1$, for all values of x. Therefore, provided $|e_0|<2$ the condition for quadratic convergence of Newton's method will be satisfied. Newton's method is usually applied without checking this condition since it converges for a very wide class of problems if x_0 is sufficiently close to the required solution. For example, taking $x_0=0.5$ gives

$$f(x_0) = 0.1065307,\ f'(x_0)=-1.10605307,$$
$$x_1 = x_0 - f(x_0)/f'(x_0) = 0.5-f(0.5)/f'(0.5) = 0.566311,$$
$$x_2 = 0.5671432, \quad (f(x_2) = 0.0000002),$$
$$x_3 = 0.5671433, \quad (f(x_3) < 0.00000001\ ).$$

This example demonstrates the rapid convergence of the Newton-Raphson iteration: in only three iterations it has been possible to find the solution correct to 6 decimal places. At each iteration it has been necessary to do more work than with the simpler linearly convergent methods, for both f and its derivative need to be calculated, but this is well compensated by the speed of convergence.

The accuracy of the computed solution depends on the condition under which the iteration is terminated. For the last example the process was halted when the difference between successive iterates was less than 10^{-7} but it is not difficult to devise examples where this is not suitable. Consider the sequence $s_N = \sum_{n=1...N} 1/n$ for which $s_{N+1}-s_N \to 0$ but $s_N \to \infty$. An alternative condition is to terminate the iteration when $f(x_n)$ is sufficiently small but if $f(x)=(1-x)^{10}$ then $x_n=1.1$ gives $f(x_n)=10^{-10}$ yet x_n is a poor approximation for the solution. In most cases the criterion used depends upon the situation in which the problem is posed but a good general purpose condition is to terminate the iteration when the relative difference $\left|(x_{N+1}-x_N)/x_{N+1}\right| < \varepsilon$ where ε is to be specified.

There still remains the outstanding problem of what to do if $f'(a) = 0$ for then Newton's method will not converge quadratically. If $f'(a) = 0$ then $x=a$ is a zero of both f and f', therefore, Newton's method is applied to f' instead of f. Having found a possible solution of $f'(x)=0$ it is substituted back into f to check if it satisfies $f(x)=0$.

Exercise 1.3.3 The secant method for solving $f(x)=0$ is obtained by approximating $f'(x_n)$ in Newton's method by $(f(x_n)-f(x_{n-1})/(x_n-x_{n-1})$ to give

$$x_{n+1}=x_n - f(x_n)(x_n-x_{n-1})/(f(x_n-f(x_{n-1})).$$

If $x_n=a+e_n$, where a is a simple zero of f, show that $e_{n+1}=Ae_ne_{n-1}$ where $A=f''(a)/[2f'(a)]$. Hence, show that $e_n=B(e_{n-1})^{\beta}$, $\beta \simeq 1.618$, i.e. the secant method converges at a rate somewhere between linear and quadratic convergence.

Exercise 1.3.4 Find all the solutions of

(i) $\sin(x) = x - 1$, x in radians,

(ii) $e^x = x + 2$.

Exercise 1.3.5 Find a value of k such that the equation $e^x=kx-2$ has a double root. Determine this double root to 5 decimal places.

Exercise 1.3.6 The function f has a root of multiplicity m at $x=\mathbf{a}$, i.e. $f(x)=(x-\mathbf{a})^m g(x)$, where $g(\mathbf{a})\neq 0$. Show that the iteration

$$x_{n+1} = x_n - m\, f(x_n)/f'(x_n)$$

converges at least quadratically to **a**. (Hint: If G(x) is the iterative function associated with Newton's method show that if f has a root of multiplicity m at $x=\mathbf{a}$ then $G'(\mathbf{a})=1-{}^1/_m$ by applying L'Hopital's rule repeatedly.)

Exercise 1.3.7 A function G(x) is said to satisfy a **Lipschitz** condition on the interval $\alpha \leqslant x \leqslant \beta$ if for all x and y in this interval there exists a constant L such that

$$|G(x) - G(y)| \leqslant L\, |x - y|.$$

Show that the iteration $x_{n+1}=G(x_n)$ converges to a solution a of $f(x)=0$, where $\alpha\leqslant\mathbf{a}\leqslant\beta$, provided the Lipschitz constant satisfies $L<1$.

1.4 Recurrence Relations Involving Matrices

We begin this section by examining an alternative way of finding the solution of a two point boundary value recurrence relation problem. In Exercise 1.2.3 the possibility of induced instability in such a problem was demonstrated; we now give an alternative method which does not suffer from this handicap.

Consider the solution of the second order linear recurrence relation

$$y_{n+1} + a_n y_n + b_n y_{n-1} = c_n, \quad 0 < n < N,$$

subject to the boundary conditions that $y_0=\alpha$ and $y_N=\beta$. We write out all the equations for n=1,...,N-1 and then determine the solution of the problem by solving the system of linear equations given by

$$\begin{aligned} b_1 y_0 + a_1 y_1 + y_2 \quad &= c_1 \\ b_2 y_1 + a_2 y_2 + y_3 \quad &= c_2 \\ \vdots \quad & \\ \vdots \quad & \\ \vdots \quad & \\ b_{N-1} y_{N-2} + a_{N-1} y_{N-1} + y_N &= c_{N-1}. \end{aligned} \tag{1.4.1}$$

These equations can be written in the matrix form

$$\begin{bmatrix} a_1 & 1 & 0 & \cdots & & & 0 \\ b_2 & a_2 & 1 & 0 & \cdots & & 0 \\ 0 & b_3 & a_3 & 1 & 0 & \cdots & 0 \\ & & & \vdots & & & \\ & & & \vdots & & & \\ & & & \vdots & & & \\ 0 & 0 & \cdots & 0 & b_{N-2} & a_{N-2} & 1 \\ 0 & 0 & \cdots & & 0 & b_{N-1} & a_{N-1} \end{bmatrix} \begin{bmatrix} y_1 \\ y_2 \\ y_3 \\ \\ \\ \\ y_{N-2} \\ y_{N-1} \end{bmatrix} = \begin{bmatrix} c_1 - b_1\alpha \\ c_2 \\ c_3 \\ \\ \\ \\ c_{N-2} \\ c_{N-1} - \beta \end{bmatrix}$$

or more briefly as **Ax**=**b**, where the ith component of the vector x is y_i, i=1,2,...,N-1, and A and b are as given above. For all but the very simplest problems it is necessary to use some form of computer to produce the solution, in which case there may possibly be difficulties in representing the coefficients exactly, for example, it may be necessary to round terms to whatever precision is used. Whether these errors and any introduced by the method of solution are significant needs careful examination. In the next section we will consider some of these difficulties.

1.4.1 The Conditioning of the Linear Equations Ax=b

To illustrate the problems inherent in solving a system of n linear equations

$$\mathbf{Ax} = \mathbf{b}, \qquad (1.4.2)$$

we consider the situation where the right hand side **b** may not be known exactly. Due to uncertainty in the elements of **b** or to rounding error in representing them we suppose that instead we determine a vector $\mathbf{x}^*$ which satisfies

$$\mathbf{Ax}^* = \mathbf{b} + \delta\mathbf{b}, \qquad (1.4.3)$$

where $\delta\mathbf{b}$ is the error in **b**. Subtracting equation (1.4.3) from (1.4.2) gives $\mathbf{A}(\mathbf{x}^*-\mathbf{x}) = \delta\mathbf{b}$. If $\delta\mathbf{b}$ is "small" then we would like the difference $\delta\mathbf{x}=(\mathbf{x}^*-\mathbf{x})$ to be "small", if not then the problem is inherently ill-conditioned. It is now necessary to decide what we mean by saying that a vector is "small" and to do this we introduce the idea of a vector norm.

Definition 1.4.1 For any n-dimensional vector **v** we define a norm of **v**, written $\|\mathbf{v}\|$, to be a real number which satisfies

(i) $\|\mathbf{v}\| \geq 0$, $\|\mathbf{v}\| = 0$ if and only if $\mathbf{v}=\mathbf{0}$,
(ii) For any scalar α and any vector **v**, $\|\alpha\mathbf{v}\| = |\alpha|\,\|\mathbf{v}\|$,
(iii) For any two vectors **v** and **w**
$\|\mathbf{v} + \mathbf{w}\| \leq \|\mathbf{v}\| + \|\mathbf{w}\|$. (The triangle inequality).

Examples 1.4.1 If $\mathbf{v} = (v_1, v_2, \dots v_n)^T$ then

(i) $\|\mathbf{v}\|_2 = \left(\sum_{i=1..n} |v_i|^2\right)^{1/2}$,

(ii) $\|\mathbf{v}\|_1 = |v_1| + |v_2| + \dots + |v_n|$,

(iii) $\|\mathbf{v}\|_\infty = \max_{1\leq i\leq n} |v_i|$.

Definition 1.4.2 For any matrix A we define the matrix norm, $\|A\|$, subordinate to a given vector norm to be

$$\|A\| = \sup_{\|v\|=1} \|Av\|.$$

Theorem 1.4.1 A subordinate norm of a matrix A satisfies the following:-

(i) $\|A\| \geqslant 0$, $\|A\| = 0$ if and only if $A = 0$.

(ii) For any scalar α and any matrix A, $\|\alpha A\| = |\alpha|\|A\|$.

(iii) For any two matrices A and B,

$$\|A + B\| \leqslant \|A\| + \|B\|,$$

$$\|AB\| \leqslant \|A\| \; \|B\|.$$

(iv) For any n dimensional vector v and any matrix A,

$$\|Av\| \leqslant \|A\| \; \|v\|.$$

We may also define matrix norms which are not necessarily subordinate to an apparently corresponding vector norm. For example, the Euclidean or Schur norm, $\|A\|_E = [\sum_{i,j=1..n} |a_{ij}|^2]^{1/2}$, is not subordinate to $\|v\|_2$ as might be expected. The norm subordinate to $\|v\|_2$ is called the spectral norm and is given by

$$\|A\|_2 = \sqrt{[\text{maximum eigenvalue of } A^H A]}$$

where A^H is the hermitian form of A, i.e. the transpose of the conjugate of A. (See Appendix 1 for a brief summary of the algebraic eigenvalue problem.)

We now return to the relationship between $\|\delta x\|$ and $\|\delta b\|$ and note that

$$\|\delta x\| = \|A^{-1}\delta b\| \leqslant \|A^{-1}\| \; \|\delta b\|.$$

This shows that an error δb may be amplified by as much as $\|A^{-1}\|$. However, it is the relative error which is more important, for a large absolute error in x may be of little consequence if x itself is very large. Rearranging $\|\mathbf{b}\|=\|\mathbf{Ax}\| \leq \|\mathbf{A}\|\,\|\mathbf{x}\|$ as $1/\|\mathbf{x}\| \leq \|\mathbf{A}\|/\|\mathbf{b}\|$ gives

$$\|\delta\mathbf{x}\|/\|\mathbf{x}\| \leq k(\mathbf{A})\,\|\delta\mathbf{b}\|/\|\mathbf{b}\|,$$

where $k(\mathbf{A})=\|\mathbf{A}^{-1}\|\,\|\mathbf{A}\|$ is called the condition number of A with respect to linear equations. If the condition number of a matrix is small then the equations **Ax=b** are said to be well conditioned. The larger the condition number is the more ill-conditioned the equations may be. This ill-conditioning is inherent within the problem and is independent of the method which is used to produce a solution.

Example 1.4.2 The Hilbert matrix of order n is given by

$$H_n = \begin{bmatrix} 1/2 & 1/3 & 1/4 & & 1/(n+1) \\ 1/3 & 1/4 & 1/5 & & \\ 1/4 & 1/5 & & & \\ & & \vdots & & \\ & & \vdots & & \\ 1/(n+1) & & & & 1/2n \end{bmatrix}$$

Find the condition number $k(H_4)$.

Solution. We have

$$H_4 = \begin{bmatrix} 1/2 & 1/3 & 1/4 & 1/5 \\ 1/3 & 1/4 & 1/5 & 1/6 \\ 1/4 & 1/5 & 1/6 & 1/7 \\ 1/5 & 1/6 & 1/7 & 1/8 \end{bmatrix} \qquad H_4^{-1} = \begin{bmatrix} 200 & -1200 & 2100 & -1120 \\ -1200 & 8100 & -15120 & 8400 \\ 2100 & -15120 & 29400 & -16800 \\ -1120 & 8400 & -16800 & 9800 \end{bmatrix}$$

therefore,

$$\|H_4\|_1=\|H_4\|_\infty = 77/60, \quad \|H_4^{-1}\|_1=\|H_4^{-1}\|_\infty = 63420$$

giving $k(H_4) = 81{,}389$ and so the solution of $H_4\mathbf{x} = \mathbf{b}$ is subject to ill-conditioning. As n increases this inherent ill-conditioning becomes worse, for example, $k(H_6) \simeq 2.7\times10^7$.

In order to illustrate the effect of a large condition number consider the solution of $H_4\mathbf{x}=\mathbf{b}$ and the perturbed equations $H_4\mathbf{x}^*=\mathbf{b}+\delta\mathbf{b}$ where $\delta\mathbf{b}=(0.01,-0.01,0.01,-0.01)^T$. The difference between x and x^* is given by $\mathbf{x}-\mathbf{x}^* = H_4^{-1}\delta\mathbf{b}$, which can be evaluated, using the inverse given in Example 1.4.2, to give $(46.2,-328.2,634.2,-361.2)^T$. We see that the original perturbation in the third component of $\delta\mathbf{b}$ has been amplified by 63420, demonstrating the sensitivity of the problem to perturbations of this kind.

Exercise 1.4.1 Consider the solution of the equations

$$10^4x + y = 10^4$$
$$x + y = 1.$$

Show that the condition number of the matrix associated with these equations is approximately 10^4. Show that by dividing the first equation by a factor of 10^4 the condition number is reduced to approximately 2 and the equations become relatively insensitive to perturbations. This process is called scaling and attempts to make any dominant element of order 1. It can often reduce the condition number of a matrix but is not guaranteed to work as Example 1.4.1 demonstrates.

It is also possible to consider the effect upon the solution of $\mathbf{Ax}=\mathbf{b}$ caused by perturbations of A. For example, let us suppose that $\mathbf{A}$ is replaced by $A+\delta A$ or $A(I+\delta B)$, where $\delta A=A\delta B$, then the computed solution $x+\delta\mathbf{x}$ satisfies $A(I+\delta B)(\mathbf{x}+\delta\mathbf{x}) = \mathbf{b}+\delta\mathbf{b}$. Expanding this expression and ignoring the second order term, $\delta B\,\delta\mathbf{x}$, gives $\mathbf{A}(I+\delta B)\,\delta x=\delta\mathbf{b}$, therefore, the relative error caused by such perturbations satisfies

$$\|\delta\mathbf{x}\|/\|\mathbf{x}\| \leqslant k(\mathbf{A})\ \|I+\delta B\|\ \|\delta\mathbf{b}\|/\|\mathbf{b}\|.$$

For full details see Wilkinson (1965).

1.4.2 Gaussian Elimination and Partial Pivoting

The standard method of solving the system of n linear equations **Ax=b** is to reduce A to a simpler form, usually upper triangular, by adding multiples of one row to the others. For example, given the equations

$$\begin{aligned}
a_{11}x_1+a_{12}x_2+\dots+ a_{1n}x_n &= b_1\\
a_{21}x_1+a_{22}x_2+\dots+ a_{2n}x_n &= b_2\\
&\vdots\\
a_{n1}x_1+a_{n2}x_2+\dots+ a_{nn}x_n &= b_n
\end{aligned} \qquad (1.4.4)$$

the terms a_{21}, $a_{31}, \dots a_{n1}$ can be eliminated by adding $-a_{j1}/a_{11}$ times the 1st equation to the jth, giving

$$\begin{aligned}
a_{11}x_1 + a_{12}x_2 + \quad \dots \quad + a_{1n}\, x_n &= b_1\\
a'_{22}x_2 + \quad \dots \quad + a'_{2n}x_n &= b'_2\\
&\vdots\\
a'_{n2}x_2 + \quad \dots \quad + a'_{nn}x_n &= b'_n
\end{aligned}$$

where the ' distinguishes the new values produced by this operation. Similarly the entries beneath the diagonal in the second and subsequent columns are eliminated until the equations become the upper triangular form:-

$$\begin{aligned}
a_{11}x_1+a_{12}x_2+ \quad \dots \quad +a_{1n}\, x_n &= b_1\\
a'_{22}x_2+ \quad \dots \quad +a'_{2n}x_n &= b'_2\\
&\vdots\\
a''_{n-1n-1}x_{n-1}+a''_{n-1n}x_n &= b''_{n-1}\\
a''_{nn}x_n &= b''_n \quad .
\end{aligned}$$

This process is called Gaussian Elimination. Having found the upper triangular form of the equations the solution is simply obtained by a backward substitution, i.e.

$$x_n = b_n/a_{nn},$$
$$x_{n-1} = (b_{n-1} - a_{n-1n}x_n)/a_{n-1n-1},$$
$$\vdots$$
$$\vdots$$
$$x_1 = (b_1 - \sum_{j=2..n} a_{1j}x_j)/a_{11},$$

where the primes have been removed for clarity but the elements given are those which result after the elimination process. It is clear that if any of the multipliers, e.g. $-a_{ji}/a_{ii}$, exceed 1 in absolute value then any error produced at any stage of this decomposition will be amplified. It is possible to show, see Wilkinson (1965), that if no multiplier exceeds 1 in absolute value then the Gaussian Elimination is stable. The computed solution x^* then satisfies

$$(\mathbf{A} + \delta\mathbf{A})\ x^* = b + \delta\mathbf{b},$$

where the maximum elements of $\delta\mathbf{A}$ and $\delta\mathbf{b}$ cannot exceed $^{n\varepsilon}/_2$, ε being the precision used in the computation. We can ensure that no multiplier exceeds 1 in absolute value by a process called **partial pivoting.**

The order in which the equations (1.4.4) appear is immaterial and they may be permuted without changing the solution. Partial pivoting is the procedure whereby before computing the multipliers we examine the pivotal column, i.e. the elements on or beneath the diagonal of the jth column at the jth step of the elimination process, and re-order the equations so that the largest element lies on the diagonal. For example, if we arrange that a_{11} is the largest element amongst (a_{i1}), i = 1, 2, .. ,n, then $|a_{i1}/a_{11}| \leqslant 1$, for each i, similarly, for the entries beneath the diagonal of the second and succeeding columns. In this way all the multipliers will have modulus $\leqslant 1$.

1.4.3 Tridiagonal Matrices

For the most part we shall be interested in very special matrices which have a particular structure, namely matrices which are **tridiagonal.** These matrices only have elements on (i) the main diagonal, (ii) one diagonal above and (iii) one diagonal below the main diagonal, see for example equations (1.4.1). Matrices of this type will appear frequently in later chapters and so we examine how to obtain the solution of tridiagonal equations. We outline a process called the LU decomposition which, though it may be used for a general matrix, has many attractions for special forms such as tridiagonal matrices.

Let A be the tridiagonal n x n matrix

$$A = \begin{bmatrix} b_1 & c_1 & & & & \\ a_2 & b_2 & c_2 & & & \\ & \cdot & \cdot & \cdot & & \\ & & \cdot & \cdot & \cdot & \\ & & & \cdot & \cdot & \cdot \\ & & & a_{n-1} & b_{n-1} & c_{n-1} \\ & & & & a_n & b_n \end{bmatrix} \qquad (1.4.5)$$

then A may be decomposed into the product of a lower triangular matrix L and an upper triangular matrix U. In general this decomposition is not unique. However, if all the diagonal entries of **L** are set to 1 then there is only one such decomposition, i.e.

$$L = \begin{bmatrix} 1 & & & & \\ l_2 & 1 & & & \\ & \cdot & \cdot & & \\ & & \cdot & \cdot & \\ & & & \cdot & \cdot \\ & & & l_n & 1 \end{bmatrix}, \quad U = \begin{bmatrix} u_1 & v_1 & & & \\ & u_2 & v_2 & & \\ & & \cdot & \cdot & \\ & & & \cdot & \cdot \\ & & & & \cdot \\ & & & & u_n \end{bmatrix}.$$

Multiplying out the product **LU** and comparing the result with A gives

$$u_1 = b_1, \qquad v_1 = c_1,$$
$$l_j u_{j-1} = a_j, \quad l_j v_{j-1} + u_j = b_j, \; j=2,3,\ldots,n. \qquad (1.4.6)$$

Therefore, the elements of the matrices L and U can be found by using the following non-linear recurrence relations,

$$u_1 = b_1, \; l_j = a_j/u_{j-1}, \; u_j = b_j - l_j c_{j-1}, \; j=2,3,\ldots,n.$$

Exercise 1.4.2 Obtain the following LU decomposition.

$$\begin{bmatrix} 5 & -1 & 0 & 0 \\ -1 & 5 & -1 & 0 \\ 0 & -1 & 5 & -1 \\ 0 & 0 & -1 & 5 \end{bmatrix} = \begin{bmatrix} 1 & 0 & 0 & 0 \\ -\frac{1}{5} & 1 & 0 & 0 \\ 0 & -\frac{5}{24} & 1 & 0 \\ 0 & 0 & -\frac{24}{115} & 1 \end{bmatrix} \begin{bmatrix} 5 & -1 & 0 & 0 \\ 0 & \frac{24}{5} & -1 & 0 \\ 0 & 0 & \frac{115}{24} & -1 \\ 0 & 0 & 0 & \frac{551}{115} \end{bmatrix}$$

The solution of the linear equations **Ax**=**b** can then be found by introducing the intermediate solution z as follows. Let **Uz**=**b** then **Ax** = **LUx** = **Lz** = **b**. We solve **Lz** = **b** by a forward substitution and then obtain x by solving **Ux** = z by a backward substitution.

Example 1.4.3 Solve the system of equations

$$\begin{bmatrix} 5 & -1 & 0 & 0 \\ -1 & 5 & -1 & 0 \\ 0 & -1 & 5 & -1 \\ 0 & 0 & -1 & 5 \end{bmatrix} \begin{bmatrix} x_1 \\ x_2 \\ x_3 \\ x_4 \end{bmatrix} = \begin{bmatrix} 4.3 \\ 3.8 \\ 3.1 \\ 4.9 \end{bmatrix}.$$

Solution. We define a vector z such that Lz=**b**, then

$$\begin{bmatrix} 1 & 0 & 0 & 0 \\ -\frac{1}{5} & 1 & 0 & 0 \\ 0 & -\frac{5}{24} & 1 & 0 \\ 0 & 0 & -\frac{24}{115} & 1 \end{bmatrix} \begin{bmatrix} z_1 \\ z_2 \\ z_3 \\ z_4 \end{bmatrix} = \begin{bmatrix} 4.3 \\ 3.8 \\ 3.1 \\ 4.9 \end{bmatrix}.$$

By using a forward substitution the components of **z** are given by

$$\begin{aligned} z_1 &= 4.3, \\ -z_1/5 + z_2 &= 3.8, \text{ or } z_2=4.66, \\ -5z_2/24 + z_3 &= 3.1, \text{ or } z_3=4.07083333, \\ -24z_3/115 + z_4 &= 4.9, \text{ or } z_4=5.74956522. \end{aligned}$$

The solution **x** is then given by solving **Ux=z**, i.e.

$$\begin{bmatrix} 5 & -1 & 0 & 0 \\ 0 & \frac{24}{5} & -1 & 0 \\ 0 & 0 & \frac{115}{24} & -1 \\ 0 & 0 & 0 & \frac{551}{115} \end{bmatrix} \begin{bmatrix} x_1 \\ x_2 \\ x_3 \\ x_4 \end{bmatrix} = \begin{bmatrix} z_1 \\ z_2 \\ z_3 \\ z_4 \end{bmatrix}$$

or, by backward substitution,

$$\begin{aligned} x_4 &= 115z_4/551 = 1.2, \\ 115x_3/24 - x_4 &= z_3, \quad x_3=1.1, \\ 24x_2/5 - x_3 &= z_2, \quad x_2=1.2, \\ 5x_1 - x_2 &= z_1, \quad x_1=1.1. \end{aligned}$$

The matrix **U** is the upper-triangular form to which A would be reduced by Gaussian Elimination, whilst the elements of **-L** are the multipliers which would have been used. Therefore, for a stable process we require that all the entries of L satisfy $|l_i| \leq 1$, $i=2,3,..,n$. We are now unable to carry out partial pivoting without destroying the tridiagonal form of A but if we can determine conditions which guarantee that the entries of L are less than 1 in absolute value then partial pivoting becomes unnecessary.

Definition 1.4.3 An n×n matrix A is **diagonally dominant** if

$$|a_{ii}| > \sum_{j=1, j\neq i}^{j=n} |a_{ij}|, \qquad i=1,2,...,n.$$

Definition 1.4.4 An n×n matrix A is **strictly diagonally dominant** if

$$|a_{ii}| > \sum_{j=1,j\neq i}^{j=n} |a_{ij}|, \qquad i=1,2,\dots,n.$$

The stability of the **LU** decomposition of **A** depends on the diagonal dominance of the transpose of A. For a tridiagonal matrix it is fairly simple to establish this result.

Theorem 1.4.2 If the matrix A given by equation (1.4.5) has a diagonally dominant transpose then all the elements of the lower triangular matrix L in its LU decomposition have modulus less than or equal to 1.

Proof. As A^T is diagonally dominant $|a_2| \leqslant |b_1|$, therefore, from equations (1.4.6)

$$|l_2| = |a_2/b_1| \leqslant 1.$$

The proof that $|l_j| \leqslant 1$, j=2,3,...,n, follows by induction. We assume that $|l_{m-1}| \leqslant 1$, for some m, and then note that

$$|l_{m-1}c_{m-2}| \leqslant |c_{m-2}|.$$

Therefore,

$$|b_{m-1}| - |l_{m-1}c_{m-2}| \geqslant |b_{m-1}| - |c_{m-2}| \geqslant |a_m|$$

and so

$$|l_m| = |a_m/(b_{m-1} - l_{m-1}c_{m-2})| \leqslant 1,$$

using the diagonal dominance of A^T, i.e. $|b_{m-1}| \geqslant |a_m| + |c_{m-2}|$, which is the required result.

Corollary 1.4.1 Let A be an n×n matrix given by equation (1.4.5) then the solution of the linear equations **Ax=b** using the LU decomposition is stable without partial pivoting provided A^T is diagonally dominant.

1.4.4 Iterative Refinement

In this section we indicate how it is possible, in general, to improve upon a computed solution of the linear equations $\mathbf{Ax} = b$ using a process called Iterative Refinement. Let us assume that by some means we have computed an approximation for the solution, which we denote by x_0, then $A\mathbf{x}_0 - b$ should be zero, but because of the errors which are inevitably introduced we shall have $\mathbf{Ax}_0 - \mathbf{b} = \mathbf{r}_0$. It is necessary to determine the residual vector, $\mathbf{r}_0$, using multiple length arithmetic otherwise it may be subject to cancellation errors and a subsequent loss of accuracy. Now define e_0 to be the solution of $\mathbf{Ae}_0 = \mathbf{r}_0$, then $A(\mathbf{x}_0 - e_0) = b$ and so the exact solution of $\mathbf{Ax} = b$ is given by $x_0 - e_0$. However, when finding e_0 further approximations will have to be made and so $x_1 = \mathbf{x}_0 - \mathbf{e}_0$ will again only be an approximation for the solution x; but it should be better than $\mathbf{x}_0$. This process can then be repeated to find $r_1 = \mathbf{Ax}_1 - \mathbf{b}$ and so on. Notice that in order to find e_0 it is necessary to reduce A to upper triangular form exactly as when finding x_0, and likewise for e_1, so it is necessary to retain the reduced form of A and also how it was obtained, or at least any permutation of the rows which took place during any partial pivoting. In the case of an LU decomposition it is necessary to retain both L and U. Further details and a complete error analysis are given by Wilkinson (1965).

Exercise 1.4.3 Given that $(1,1,1,1)^T$ is an approximate solution of the linear equations

$$\begin{bmatrix} 5 & -1 & 0 & 0 \\ -1 & 5 & -1 & 0 \\ 0 & -1 & 5 & -1 \\ 0 & 0 & -1 & 5 \end{bmatrix} \begin{bmatrix} x_1 \\ x_2 \\ x_3 \\ x_4 \end{bmatrix} = \begin{bmatrix} 4.3 \\ 3.8 \\ 3.1 \\ 4.9 \end{bmatrix},$$

use the iterative refinement method to compute a more accurate solution.

1.4.5 The Iterative Solution of Linear Equations

Section 1.4.4 introduced an iterative method for the solution of the n linear equations **Ax=b**; we now consider other such methods. Let **L** be the lower triangular matrix consisting of all the elements beneath the diagonal of the matrix A, let D be the diagonal matrix which consists of the diagonal elements of A and let U the upper triangular matrix made up from the rest of A, then A=L+D+U. Now write Ax=(L+D+U)x=b as Dx=-(L+U)x + b or equivalently

$$x = D^{-1} [-(L+U)x + b]$$

which can be turned into a recurrence relation for a vector x_r by putting

$$x_{r+1} = D^{-1} [-(L+U)x_r + b] = Gx_r + c,$$

where $G=-D^{-1}[L+U]$ and $c=D^{-1}b$. Notice that $x=Gx+c$. This iteration is called Jacobi's method and given a vector x_0 a sequence of vectors, x_r, can be generated. It is easy to show that if $\|G\| < 1$ this sequence converges to the solution of **Ax=b.** Consider the difference

$$x_{r+1} - x = Gx_r + c - Gx - c = G(x_r-x)$$

then

$$\|x_{r+1}-x\| \leqslant \|G\| \; \|x_r-x\| \leqslant \|G\|^2 \|x_{r-1}-x\|$$
$$\dots\dots \leqslant \|G\|^{r+1} \|x_0-x\|.$$

Therefore, if $\|G\| < 1$, the difference $x_{r+1} \to x$ as $r\to\infty$. **Although Jacobi's method is developed in terms of a matrix equation it is best applied in the component form:-**

$$[x_{r+1}]_i = \{ -\sum_{i\neq j} a_{ij}[x_r]_i + b_i \}/ a_{ii}, \qquad i=1,2,..,n$$

where $[x_r]_i$ is the i-th component of the vector x_r. Furthermore, it is not necessary to keep every member of the sequence of vectors $\{x_r\}$, r=1,2,....; once x_{r+1} has been computed x_r can be discarded. This means that only two vectors are required, one contains the current solution, x_r, and the second contains the next member of the sequence, x_{r+1}. Once x_{r+1} has been calculated x_r can be overwritten by x_{r+1} and the process repeated.

Exercise 1.4.4 Show that if the matrix A is strictly diagonally dominant then Jacobi's method converges.

In order to compute the jth element of the vector x_{r+1}, $[x_{r+1}]_j$, using the Jacobi process it is necessary to use the values $[x_r]_1$, $[x_r]_2$, ..., $[x_r]_{j-1}$, $[x_r]_{j+1}$,...,$[x_r]_n$. However, a better value for $[x_{r+1}]_j$ can be found if we use $[x_{r+1}]_1$, ..., $[x_{r+1}]_{j-1}$, $[x_r]_{j+1}$, ..., $[x_r]_n$ since values for $[x_{r+1}]_1$,..,$[x_{r+1}]_{j-1}$ will already have been found. This process can be written in the component form

$$[x_{r+1}]_i = \{-\sum_{j<i} a_{ij}[x_{r+1}]_j - \sum_{j>i} a_{ij}[x_r]_j + b_i\}/a_{ii}, \qquad i=1,2,\ldots,n$$

which is called the Gauss-Seidel method. Let us suppose that some vector v contains the current solution, then in order to update the jth element we use the above expression and then overwrite v_j immediately; therefore, only one vector is required to determine the solution.

Exercise 1.4.5 Show that the above equations can be written in the matrix form

$$\mathbf{x}_{r+1} = -(D+L)^{-1}[\,U\mathbf{x}_r - b].$$

Conditions under which the Jacobi and Gauss-Seidel methods converge are beyond the scope of this text, however, they are frequently applied in the component forms above and it is usually obvious, after a number of iterations, whether or not they are going to converge. Under relatively simple conditions it is possible to show that either both methods converge or both diverge, and that when they do converge then the Gauss-Seidel method does so more rapidly. (See Varga (1962), Young (1971).)

Example 1.4.4 Use Jacobi's and the Gauss-Seidel method to find a solution of

$$\begin{bmatrix} 5 & -1 & 0 & 0 \\ -1 & 5 & -1 & 0 \\ 0 & -1 & 5 & -1 \\ 0 & 0 & -1 & 5 \end{bmatrix} \begin{bmatrix} x_1 \\ x_2 \\ x_3 \\ x_4 \end{bmatrix} = \begin{bmatrix} 4.3 \\ 3.8 \\ 3.1 \\ 4.9 \end{bmatrix}$$

taking $x_0=(1,1,1,1)^T$. (The solution is $(1.1,1.2,1.1,1.2)^T$.)

Solution. Let $[x_r]_i$ denote the i-th component of the r-th approximation for the solution x and set $[x_o]_i=1$, i=1,2,3,4. Then the first equation gives $5x_1-x_2=4.3$ and so

$$5[x_1]_1 = 4.3 + [x_o]_2 \quad \text{or} \quad [x_1]_1 = (4.3 + [x_o]_2)/5 = 1.06,$$

The second and subsequent equations give:-

$$5[x_1]_2 = 3.8 + [x_o]_1 + [x_o]_3] \text{ and so } [x_1]_2=1.16,$$
$$5[x_1]_3 = 3.1 + [x_o]_2 + [x_o]_4 \text{ and so } [x_1]_3=1.02,$$
$$5[x_1]_4 = 4.9 + [x_o]_3 \text{ and so } [x_1]_4=1.18.$$

The iteration is then repeated to find $[x_2]_i$, i=1,2,3,4, and so on. The results of Jacobi's method can be summarised by the following table.

Table 1.4.1. Jacobi's Method applied to Example 1.4.4

r	$[x_r]_1$	$[x_r]_2$	$[x_r]_3$	$[x_r]_4$
1	1.06	1.16	1.02	1.18
2	1.092	1.176	1.088	1.184
3	1.0952	1.196	1.092	1.1976
4	1.0992	1.19744	1.09872	1.1984
:	:	:	:	:
10	1.099998	1.199997	1.099998	1.199998

Likewise for the Gauss-Seidel method we obtain the following table.

Table 1.4.2. Gauss-Seidel Method applied to Example 1.4.4

r	$[x_r]_1$	$[x_r]_2$	$[x_r]_3$	$[x_r]_4$
1	1.06	1.172	1.0844	1.19088
2	1.0944	1.18976	1.096128	1.1992256
3	1.097952	1.198816	1.099608	1.1999217
4	1.099652	1.199874	1.099959	1.1999920
:	:	:	:	:
10	1.100000	1.200000	1.100000	1.2000000

This demonstrates that the Gauss-Seidel method does indeed converge faster than Jacobi's method for this example.

1.4.6 Recurrence Methods for the Determination of Eigenvalues

Let A be an n×n matrix and x a non-zero vector then x is called an eigenvector of A if there exists a scalar λ such that $A\mathbf{x} = \lambda\mathbf{x}$. The scalar λ is called the eigenvalue associated with x. The determination of some or all the eigenvectors and eigenvalues associated with a matrix is a problem which frequently arises. A brief summary of this problem is given in Appendix 1. Finding even a single eigenvalue is far from simple but it is sometimes possible to determine the modulus of largest eigenvalue by using the so called Power Method. We shall only consider matrices which are real and symmetric since this guarantees that all the eigenvalues are real. Let w_0 be any arbitrary vector and then define the vector recurrence

$$\mathbf{v}_{r+1} = A\mathbf{w}_r, \quad \mathbf{w}_{r+1} = (\mathbf{v}_{r+1})/\|\mathbf{v}_{r+1}\|. \qquad (1.4.7)$$

Clearly, $\mathbf{w}_r=(A\mathbf{w}_{r-1})/\|A\mathbf{w}_{r-1}\|=(A^2\mathbf{w}_{r-2})/\|A^2\mathbf{w}_{r-2}\| = (A^r\mathbf{w}_0)/\|A^r\mathbf{w}_0\|$. If A has n distinct eigenvalues such that $|\lambda_1|>|\lambda_2|>\ldots.>|\lambda_n|$ then it is possible to expand w_0 in terms of a system of linearly independent normalised eigenvectors, $A\mathbf{u}_i = \lambda_i\mathbf{u}_i$, i=1,..,n, to give

$$\mathbf{w}_0 = \sum_{i=1..n} \alpha_i \mathbf{u}_i .$$

Provided $\alpha_1 \neq 0$

$$A^r\mathbf{w}_0 = \sum_{i=1} \alpha_i A^r \mathbf{u}_i = \sum_{i=1} \alpha_i \lambda_i{}^r \mathbf{u}_i$$

$$= \lambda_1{}^r [\ \alpha_1 u_1 + \sum_{i=2} \alpha_i (\lambda_i/\lambda_1)^r \mathbf{u}_i\ ]$$

$$\to \lambda_1{}^r \alpha_1 u_1 ,\ \text{as}\ r\to\infty.$$

In which case $\|A^r\mathbf{w}_0\| \to |\lambda_1 \alpha_1| \|u\| = |\lambda_1{}^r \alpha_1|$ therefore $w_r \to u_1$. Similarly $v_{r+1}=\mathbf{A}\mathbf{w}_r \to \lambda_1 u_1$ and $\|v_{r+1}\| \to |\lambda_1|$. This gives a process which determines the eigenvalue of maximum modulus and the corresponding eigenvector of a given matrix A. However, if $|\lambda_1/\lambda_2|$ is close to 1 then the rate of convergence will be poor. If $|\lambda_1|=|\lambda_2|=\ldots=|\lambda_m|$ and $|\lambda_1|>|\lambda_{m+1}|> \ldots >|\lambda_n|$ then this method will still converge, at a reduced rate, and although we obtain the dominant eigenvalue the corresponding eigenvectors are not so easily found. This method can be extended to more general problems; full details are given by Wilkinson (1965).

Example 1.4.5 Find the spectral norms of the matrices associated with the Jacobi and Gauss-Seidel methods in Example 1.4.4.

Solution. For Jacobi's method the iterative matrix, $J=-D^{-1}(L+U)$, is given by

$$J = \begin{bmatrix} 0 & 0.2 & 0 & 0 \\ 0.2 & 0 & 0.2 & 0 \\ 0 & 0.2 & 0 & 0.2 \\ 0 & 0 & 0.2 & 0 \end{bmatrix}$$

Since J is real and symmetric $\|J\|_2=|\text{maximum eigenvalue of } J|$. If we take $w_0=(1,1,1,1)^T$ the recurrence relations (1.4.7) give $\mathbf{v}_1=\mathbf{J}\mathbf{w}_0=(0.2,0.4,0.4,0.2)^T$, $\|v_1\|_2=0.632455$, and $\|v_2\|=0.3224903$, ... $\|v_{10}\|=0.323606$ which gives the largest eigenvalue of J correct to six decimal places. Since $\|J\|<1$ Jacobi's method will converge for the equations given in Example 1.4.4 from any initial estimate.

For the Gauss-Seidel method the iterative matrix is given by

$$\mathbf{G} = -(D+L)^{-1}U = \begin{bmatrix} 0 & \alpha & 0 & 0 \\ 0 & \alpha^2 & \alpha & 0 \\ 0 & \alpha^3 & \alpha^2 & \alpha \\ 0 & \alpha^4 & \alpha^3 & \alpha^2 \end{bmatrix}, \qquad \alpha = 0.2.$$

This matrix is not symmetric but applying the Power Method to $G^H\mathbf{G}$, with $\mathbf{w}_0=(1,1,1,1)^T$, gives $\|\mathbf{v}_9\|=\|\mathbf{v}_{10}\|=0.0109662$ and so $\|\mathbf{G}\|_2 \simeq 0.10472$. Notice that $\|\mathbf{J}\|^2=\|\mathbf{G}\|$ and so, for this example, the Gauss-Seidel method will converge faster than Jacobi's method.

Exercise 1.4.6 Show that the recurrence relation given by solving the linear equations $A\mathbf{w}_{r+1}=\mathbf{v}_r$, $\mathbf{v}_{r+1}=\mathbf{w}_{r+1}/\|\mathbf{w}_{r+1}\|$ has the property that $\|\mathbf{v}_r\|$ converges to the eigenvalue of A which has the smallest modulus. (Hint: Look at the relationship between the eigenvalues of **A** and A^{-1}. Note that if A has a zero eigenvalue then det**A**=0, in which case the above method fails because the equations are singular.)

1.5 Extrapolation and Acceleration of Convergence

In the previous sections iterative methods for solving various problems were introduced in which the limit of a certain sequence was the required solution. Frequently such sequences converge only very slowly and it is necessary to accelerate them by some means. We consider now two different types of acceleration methods. The first type of method attempts to eliminate the principal term in the error between the true and computed solutions. The second type takes a convergent method and introduces a parameter; this parameter is then selected, either heuristically or analytically, to improve the rate at which the process converges. In order to illustrate both types of method we shall consider the the solution of $f(x)=0$ by a linearly convergent process.

1.5.1 Acceleration of Linearly Convergent Methods for f(x)=0

Let us recall that an iterative method, $x_{n+1}=G(x_n)$, for finding a solution of the equation $f(x)=0$ converges linearly to a solution **a** if there exists a constant C such that $|G'(x)|\leqslant C<1$, in some interval containing **a** and an initial estimate x_0. This condition is sufficient to guarantee that $x_n-\mathbf{a}\to 0$ since

$$x_{n+1}-\mathbf{a}\simeq G'(\mathbf{a})(x_n-\mathbf{a}).$$

We also have that $x_n-\mathbf{a}\simeq G'(\mathbf{a})(x_{n-1}-\mathbf{a})$ and eliminating $G'(\mathbf{a})$ between these equations gives

$$\mathbf{a}\simeq\frac{(x_{n+1}x_{n-1}-x_n^2)}{(x_{n+1}-2x_n+x_{n-1})}.$$

This procedure should give an approximation for the solution which is better than x_{n+1}, x_n or x_{n-1}. and is usually referred to as Aitken's method.

Example 1.5.1 From Example 1.3.2 the iteration $x_{n+1}=\exp(-x_n)$ converges linearly to the solution of $e^{-x}=x$ from any $x_0>0$. Taking $x_0=1$ produces $x_9=0.5648793$, $x_{10}=0.5684287$, $x_{11}=0.5664147$. The above procedure gives the improved estimate for the solution of 0.5671437 which is accurate to 5 decimal places and is a substantial improvement. This value can then be used to restart the iteration and repeat the process.

A second type of extrapolation attempts to introduce a parameter into a convergent process and then select a value to give an improved rate of convergence.

Example 1.5.2 Consider the solution of $x^2-3x+2=0$ by the iterative method $x_{n+1}=(x_n^2+2)/3$. It was shown that this iteration converged linearly, i.e. slowly, and we now seek ways to improve the rate of convergence.

Consider the alternative iteration

$$x_{n+1} + \omega x_{n+1} = (x_n^2 + 2)/3 + \omega x_n, \qquad (1.5.1)$$

where ω is a parameter. Clearly if this iteration converges to x^* then

$$x^* + \omega x^* = ((x^*)^2 + 2)/3 + \omega x^*$$

and so x^* is a solution of the quadratic $x^2-3x+2=0$. Re-arranging equation (1.5.1) gives

$$x_{n+1} = [(x_n^2 + 2)/3 + \omega x_n]/(1 + \omega),$$

so, in the notation of Section 1.3, the iterative function associated with this method is

$$G(x) = [(x^2 + 2)/3 + \omega x]/(1 + \omega).$$

This method will converge linearly to a solution, a, if $|G'(x)| < 1$ for all values of x sufficiently close to a. However, in order to obtain a higher rate of convergence it is necessary that $G'(a)=0$. Therefore, differentiating G with respect to x we have that $G'(a)=0$ provided $\omega=-2a/3$. As a is unknown we take $\omega=-2x_n/3$ giving the iterative process

$$x_{n+1} = x_n - \frac{x_n^2 - 3x_n + 2.}{2x_n - 3}$$

This is nothing more than the Newton Raphson iteration which converges quadratically provided we have a suitable value for x_0. As a bonus this accelerated iteration converges not only to the root of $x^2-3x+2=0$ at x=1 but also the second solution at x=2 provided the initial value is sufficiently close. In general there is little point in accelerating linearly convergent methods for solving f(x)=0; the Newton Raphson method is used directly. However, this type of acceleration can be very useful in other problems.

1.5.2 Acceleration Methods for Linear Equations

The relatively slow convergence of the simple Jacobi and Gauss-Seidel methods can be accelerated in several different ways. For example, the Aitken process can be applied to each component of three successive iterates but a more appropriate technique is available.

Let us suppose that some iterative process generates a sequence of approximations $\{\mathbf{x}_r\}$ for which we determine the displacements $\mathbf{d}_r=\mathbf{x}_r-\mathbf{x}_{r-1}$, then

$$\mathbf{x} = \mathbf{x}_r + (\mathbf{x}_{r+1}-\mathbf{x}_r) + (\mathbf{x}_{r+2}-\mathbf{x}_{r+1}) + \ldots$$
$$= \mathbf{x}_r + \mathbf{d}_{r+1} + \mathbf{d}_{r+2} + \ldots\ldots$$

It is frequently observed that as r gets larger i.e. the ratio of the i-th component of the vector d_{r+1} to the same component of d_r tends to the same value λ for all i, i.e. $\mathbf{d}_{r+1} \simeq \lambda \mathbf{d}_r$, in which case

$$\mathbf{x} = \mathbf{x}_r + \lambda\mathbf{d}_r + \lambda^2\mathbf{d}_r + \lambda^3\mathbf{d}_r + \ldots\ldots$$
$$=\mathbf{x}_r + \lambda(1 + \lambda + \lambda^2 + \ldots\ldots\ldots)\mathbf{d}_r = \mathbf{x}_r + \lambda\mathbf{d}_r/(1-\lambda)$$

provided $|\lambda|<1$. This is the basis of an acceleration process attributed to Lyusternik (1947). As the sequence of approximate solutions is determined a check is kept on the ratio of the elements of the displacements; if the ratio of successive elements tends to some limiting value which has absolute value less than 1 then the improved estimate given above is used. If the limit of the ratio of the elements of successive displacements has modulus greater than 1 the process will not converge. It is straightforward to show that the value λ tends to the spectral radius of the iterative matrix (as defined in Section 1.4.5) by associating this iteration with the Power Method given in Section 1.4.6.

We can also accelerate the convergence of the Gauss-Seidel method by introducing a suitable parameter. Let x_r be the r-th approximate solution of $\mathbf{Ax}=(\mathbf{L}+\mathbf{D}+\mathbf{U})\mathbf{x}=\mathbf{b}$ and $[\mathbf{x}_r]^G$ the vector that the Gauss-Seidel iteration would produce, i.e.

$$[\mathbf{x}_r]^G = -(\mathbf{D}+\mathbf{L})^{-1}(\mathbf{U}\mathbf{x}_r - \mathbf{b})$$

and then consider

$$\mathbf{x}_{r+1} = \mathbf{x}_r + \omega(\ [\mathbf{x}_r]^G - \mathbf{x}_r\ ).$$

If $\omega=1$ then $\mathbf{x}_{r+1}=[\mathbf{x}_r]^G$ and this method reduces to the standard Gauss-Seidel iteration. If $\omega>1$ then we have added slightly more to $\mathbf{x}_r$ than the correction involved with Gauss-Seidel method, accordingly this method is called the successive over-relaxation, SOR, or the extrapolated Liebmann method. From Section 1.4.5 the Gauss-Seidel method is given in component form by

$$[\mathbf{x}_{r+1}]_i=\Big(-\sum_{j<i} a_{ij}[\mathbf{x}_{r+1}]_j - \sum_{j>i} a_{ij}[\mathbf{x}_r]_j + b_i\Big)/a_{ii}. \qquad i=1,2,..,n$$

Therefore, for the SOR method we have

$$a_{ii}[\mathbf{x}_{r+1}]_i + \omega\sum_{j<i} a_{ij}[\mathbf{x}_{r+1}]_j = [(1-\omega)a_{ii}[\mathbf{x}_r]_i - \omega\sum_{j>i} a_{ij}[\mathbf{x}_r]_j + \omega b_i], \qquad i=1,2,..,n$$

or in matrix form

$$\mathbf{x}_{r+1} = (\mathbf{D}+\omega\mathbf{L})^{-1}\big[[(1-\omega)\mathbf{D} - \omega\mathbf{U}]\mathbf{x}_r + \omega\mathbf{b}\big].$$

Exercise 1.5.1 Find the solution of the problem given in Example 1.4.4 using values of ω of 1,1.025,1.05,1.075.

A necessary condition for the convergence of the SOR process is that $0<\omega<2$; but the determination of the optimum choice of ω is difficult. For certain types of matrix it can be found in terms of the spectral radius of the associated Jacobi matrix and we will return to this problem in a later chapter.

1.5.3 General Extrapolation Techniques

Many numerical methods are based upon a discretization of a continuous problem and depend upon continuity of the problem as some parameter tends to zero. If the solution of a given problem is f for which we compute an approximation f(h), depending on some parameter h, then it is sometimes possible to express the computational error as

$$f - f(h) = g_1(h) + g_2(h) + \ldots .$$

where g_1, g_2, ... are functions of h. If the functions g_i are relatively simple it is possible to eliminate them and reduce the computational error. For example, consider the problem of evaluating the definite integral

$$\int_a^b f(x)\, dx.$$

If f(x) is a positive function of x then the value of this integral is equivalent to the area bounded by the curve y=f(x), the x-axis and the lines x=a and x=b as shown in Figure 1.5.1

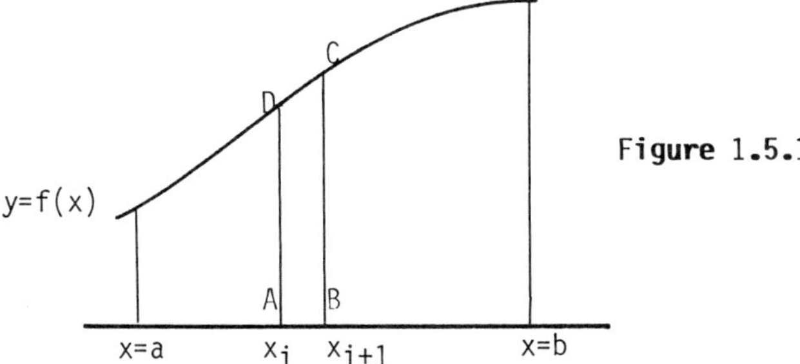

Figure 1.5.1

We begin by dividing the interval (a,b) into N sub-intervals each of width h=(b-a)/N. Then the integral over the interval (x_i, x_{i+1}), where x_i=a+ih, can be approximated by the area of the trapezium ABCD, i.e.

$$\int_{x_i}^{x_{i+1}} f(x)\, dx \simeq 0.5\ h\ \{f(x_i) + f(x_{i+1})\}. \quad (1.5.2)$$

It is possible to obtain an expression for the error associated with this approximation as follows. By the Fundamental Theorem of Calculus there exists a function F such that $F'(x)=f(x)$ and so using Taylor's Theorem repeatedly gives

$$\int_{x_i}^{x_{i+1}} F'(x)\,dx - 0.5\,h\,\{F'(x_i)+F'(x_{i+1})\}$$

$$= F(x_i+h)-F(x_i) - 0.5\,h\,\{F'(x_i)+F'(x_i+h)\}$$

$$\begin{aligned} = \; & F(x_i) + hf(x_i)+{}^{h^2}\!/_{2}f'(x_i)+{}^{h^3}\!/_{6}f''(x_i)+ \ldots\ldots \\ & - F(x_i) \\ & - {}^{h}\!/_{2}f(x_i) \\ & - {}^{h}\!/_{2}f(x_i)-{}^{h^2}\!/_{2}f'(x_i)-{}^{h^3}\!/_{4}f''(x_i) + \ldots\ldots \end{aligned}$$

$$= -\,{}^{h^3}\!/_{12}f''(x_i) + O(h^4),$$

provided the integrand, f, is sufficiently smooth. For each interval the error is of a similar form; therefore, adding together the contribution from each of the N subintervals, the overall error will be approximately $N(Ah^3 + O(h^4))$, where A is a constant independent of h but depending upon f and its various derivatives. As $Nh=(b-a)$ this gives the composite trapezium rule:-

$$\int_a^b f(x)dx = 0.5h\{f(a)+2f(a+h)+2f(a+2h)+\ldots\ldots+2f(b-h)+f(b)\} + E_T(h)$$

where $E_T(h)=O(h^2)$. A more detailed examination gives

$$E_T(h) = Ah^2 + Bh^4 + \ldots\ldots\ldots , \qquad (1.5.3)$$

where the coefficients A, B, .. are independent of h but involve the derivatives of f evaluated between x=a and x=b.

Example 1.5.3 Estimate $\int_0^1 e^{-x^2}\,dx$ using the composite trapezium rule taking N=2,4,8,16. (The value of the integral is 0.746824133 correct to 9 decimal places).

Solution. Here $f(x)=e^{-x^2}$. Consider the following values for N.

N=2,h=0.5.

$\int_0^1 f(x)\,dx \simeq 0.25\,[f(0) + 2f(0.5) + f(1)] = 0.7313702.$

N=4,h=0.25.

$\int_0^1 f(x)\,dx \simeq 0.125\,[f(0)+2f(0.25)+2f(0.5)+2f(0.75)+f(1)]=0.7429840.$

N=8,h=0.125.

$\int_0^1 f(x)\,dx \simeq 0.7458655.$

N=16,h=0.0625.

$\int_0^1 f(x)\,dx \simeq 0.7465846.$

The previous example shows that the sequence of approximations based on interval sizes h, T(h), does appear to be slowly converging to the correct value of the integral. There are other possible ways of approximating a definite integral, but even so a large number of intervals is still required.

Exercise 1.5.2 The composite Simpson's rule is given by

$$\int_a^b f(x)dx= {}^h/_6(f(a)+4f(a+h)+2f(a+2h)+4f(a+3h)+ \,.. \\ +2f(b-2h)+4f(b-h)+f(b)) + E_S,$$

where $h={}^{(b-a)}/_N$. Show that $E_S(h)=Ah^4 + Bh^6 + \ldots\ldots$.

Exercise 1.5.3 Evaluate $\int_0^1 e^{-x^2}dx$ using values of N of 2,4,8,16. Compare the result with Example 1.5.3.

It is easy to see that Simpson's rule is more accurate than the trapezium rule (1.5.2), because, for the former the principal term in the error E_S is proportional to h^4 whilst for the latter it is proportional to h^2. This means that if h is divided by 2 then the error term associated with Simpson's rule is divided by 2^4=16 whereas with the trapezium rule the error is only reduced by a factor of 2^2. By examining the error associated with the trapezium rule it is possible to show that in excess of 1200 intervals are required to evaluate the integral in Example 1.5.3 correct to 6 decimal places. This disadvantage can be overcome by examining the error term with more care. If T(h) is an approximation for some integral I then we have that

$$I - T(h) = Ah^2 + Bh^4 + .. \qquad (1.5.4)$$

and also

$$I - T(2h) = A\,(2h)^2 + B\,(2h)^4 + .. \qquad (1.5.5)$$

where A and B are constants which depend upon f and its various derivatives but not h. Subtracting (1.5.5) from 4×(1.5.4) gives

$$I - [\,4T(h) - T(2h)\,]/_3 = -4\,B\,h^4 + ...$$

that is the approximation $T_1(h)=[4T(h)-T(2h)]/_3=T(h)+[T(h)-T(2h)]/_3$ differs from the exact value of the integral by a term proportional to h^4 whereas T(h) and T(2h) differ by a term $O(h^2)$. Clearly this process can be repeated to give

$$T_2(h) = [16T_1(h) - T_1(2h)]/_{15} = T_1(h) + [T_1(h)-T_1(2h)]/_{15}. \qquad (1.5.6)$$

This process is called **Romberg extrapolation** and is applicable whenever the computational error is of the form (1.5.4). In general T_n is given by the recurrence relation

$$T_n(h) = T_{n-1}(h) + [T_{n-1}(h)-T_{n-1}(2h)]/_{(2^{2n}-1)}, \qquad (1.5.7)$$

which has error proportional to $h^{2(n+1)}$.

Exercise 1.5.4 Show that the error between $T_2(h)$ and the correct value of the integral is proportional to h^6.

Example 1.5.4 Compute an approximation for $\int_0^1 e^{-x^2}dx$ using Romberg extrapolation applied to the trapezium rule.

Solution. Using the results of Example 1.5.3 gives the following table.

Table 1.5.1. Romberg Extrapolation as applied to Example 1.5.3.

h	T(h)	$T_1(h)$	$T_2(h)$
0.5	0.7313702		
0.25	0.7429841	0.7468554	
0.125	0.7458656	0.7468261	0.7468241
0.0625	0.7465846	0.7468241	0.7468241

The value in extrapolating the convergence of the sequence T(h) is obvious.

Exercise 1.5.5 Compute an approximation for $\int_0^1 e^{-x^2}dx$ using Simpson's rule and Romberg extrapolation.

This method of eliminating the dominant term in the error is due to Richardson (1927) and is frequently used to improve the accuracy or rate at which a sequence converges. It is dependent upon knowing the form of the error but more importantly that a simple expansion of the error in terms of h exists. That this is not always the case can be easily demonstrated by considering the integral

$$\int_0^1 \sqrt{x} \log x \, dx.$$

The integrand does not have a bounded derivative at x=0 and so the expression for the error E_T given by expansion (1.5.3) is not valid. (See Fox and Mayers (1968) and Exercise 1.5.9.)

Exercise 1.5.6 If S(h) is an approximation for a definite integral I which is determined using Simpson's rule, as given in Exercise 1.5.2, show that $S_1(h) = [16S(h) - S(2h)]/_{15}$ differs from I by $O(h^6)$.

Exercise 1.5.7 Show that the first extrapolation of the trapezium rule, $T_1(h)$, gives Simpson's rule as shown in Exercise 1.5.2.

Exercise 1.5.8 Estimate the number of intervals required to evaluate $\int_0^1 e^{x^2}dx$ correct to 6 decimal places using (i) the trapezium rule, (ii) Simpson's rule. Use Romberg extrapolation to evaluate this integral to the required precision.

Exercise 1.5.9 It is possible to show that for the integral

$$I = \int_0^1 \sqrt{x} \log x \, dx$$

the error when using the trapezium rule is given by

$$I - T(h) = Ah^{3/2}\log_e h + Bh^{3/2} + Ch^2 + Dh^4 + \cdots$$

(See Fox and Mayers (1968).) Using values for h of $2^{-(r-1)}$, r=1,2,3,4, compute an accurate value for I by eliminating the principal terms in the above expression.

CHAPTER 2
Ordinary Differential Equations—Initial Value Problems

2.1 Introduction

Many mathematical models can be formulated in terms of the rate of change of one or more physical variables and naturally lead to differential equations. Furthermore, any additional conditions imposed on the model can be used to eliminate those solutions of the differential equation which are not applicable and hence determine particular solutions which are relevant. The solution of such problems can provide evidence to validate the original assumptions upon which the model was developed and may also be used to predict the consequences of these assumptions. Consider the following simple example.

Example 2.1.1 The rate of growth of the membership of a certain population is known to be proportional to the total membership. If the population grows by 10% in 1 year find an expression for the population as a function of time. How long will it take the population to double its size?

Solution. Let $p(t)$ be the total membership of the population at time t and K a constant then

$$dp/dt = Kp. \qquad (2.1.1)$$

Since this equation involves the differential coefficient of p as well as p itself it is called a **differential equation**; as there is only one independent variable, x, it is called an **ordinary differential equation.** Furthermore, as the equation (2.1.1) involves only the first derivative it is called a **first order equation.** Any function which satisfies this equation is called a **solution,** for example, $p(t) = e^{Kt}$ is a solution, but so is $p(t)=2e^{Kt}$ or $p(t)=100e^{Kt}$. In fact any function of the form $p(t)=Ae^{Kt}$, where A is an arbitrary constant, is a solution. The **general solution** of this first order equation is any solution which contains exactly one arbitrary constant, supplying a value for this constant will give a **particular solution.**

Returning to Example 2.1.1, if the population at time $t=0$ is given by P_0 then the arbitrary constant in the general solution can be found. This gives the particular solution $p(t)=P_0\exp(Kt)$. We are also given that $p(1)=1.1P_0$ and so $K=\log_e 1.1$. If the population is to double after a period of T years then $p(T)=2P_0=P_0\exp(KT)$ from which $T=\log_e 2/\log_e 1.1$ or approximately 7.27 years.

The general first order differential equation can be written as a functional relationship between an independent variable, x, a dependent variable y and its first derivative $y'=dy/dx$, i.e.

$$\phi(x,y,dy/dx) = 0. \qquad (2.1.2)$$

Any function which satisfies this expression is said to be a **solution** of the differential equation. It is sometimes possible to solve (2.1.2) for y' but this may not always be the case.

Example 2.1.2 Differential equations of the form

$$y = xy' + f(y'),\ y'={}^{dy}/_{dx},$$

are collectively known as Clairaut equations. Find all the solutions of the non-linear differential equation

$$y = xy' - (y')^2.$$

Solution. Differentiating the given equation with respect to x gives

$$y' = y' + xy'' - 2y'y''$$

which since it involves a second derivative is called a second order equation. Rearranging this expression gives

$$0 = y''(x - 2y'),$$

therefore, either

$$y'' = 0, \qquad (2.1.3)$$

or

$$x - 2y' = 0. \qquad (2.1.4)$$

Equation (2.1.3) gives that $y'=C$ and so $y(x)=Cx+D$. This cannot be the general solution since it contains two constants, C and D, but substituting back into the original differential equation we see that $Cx+D=Cx-C^2$ and so $D=-C^2$, therefore,

$$y(x)=Cx-C^2. \qquad (2.1.5)$$

This is the general solution; some of the possible particular solutions are $y(x)=x-1$, $C=1$, and $y(x)=4x-16$, $C=4$. Equation (2.1.4) implies that $y'={}^{x}/_{2}$ and so $y(x)={}^{x^2}/_{4}+B$, but this function only satisfies the differential equation when $B=0$ giving the solution $y(x)={}^{x^2}/_{4}$. Since this solution cannot be obtained by specifying a value for the arbitrary in the general solution it is said to be a **singular solution.**

The previous example demonstrates that even relatively simple differential equations may have a complicated system of solutions. Instead of considering the general form (2.1.2) we shall examine only those equations which may be solved explicitly for y', i.e.

$$y'=f(x,y). \qquad (2.1.6)$$

There is a wide class of analytical techniques which can be used to solve equations of this type but there are very many equations which do not yield to such an approach, for example, the Ricatti equation $y'=x^2+y^2$. The general solution of the differential equation (2.1.6) will involve one arbitrary constant which can be determined if the solution must satisfy an additional constraint. The determination of the solution of $y'=f(x,y)$ for $x\geqslant a$, subject to $y(a)=\alpha$, is called an **initial value problem.** The condition $y(a)=\alpha$ is called the **initial value.**

In this chapter we shall investigate various methods by which approximate solutions of initial value problems can be obtained. Although we shall not be primarily interested in the existence and uniqueness of such solutions these properties are of course essential. For example, it is easy to show that the existence of a unique solution requires more than continuity of f with respect to both x and y near $x=a$, $y=\alpha$ by considering the differential equation $y'=2\sqrt{|y|}$, subject to $y(0)=0$, which has solutions $y(x)=0$ and $y(x)=x^2$.

Definition 2.1.1 A function f(x,y) is said to be **Lipschitz** with respect to y in some closed bounded region $R\subset\mathbb{R}^2$ if for all (x,y_1) and (x,y_2) in R there exists a constant $L>0$ such that

$$|f(x,y_1) - f(x,y_2)| \leqslant L\ |y_1 - y_2|. \qquad (2.1.7)$$

The Lipschitz condition (2.1.7) is not an easy one to understand but there is a relatively simple criterion which is sufficient to ensure a function is Lipschitz within a given region. If f has a continuous partial derivative with respect to y throughout R then it is Lipschitz within this region. If $f_y = {}^{\partial f}/_{\partial y}$ exists then

$$f(x,y_1) - f(x,y_2) = (y_1 - y_2)\, f_y(x,\theta),$$

for some θ between y_1 and y_2, and if ${}^{\partial f}/_{\partial y}$ is continuous in R then it is bounded and hence (2.1.7) is satisfied. However, if ${}^{\partial f}/_{\partial y}$ is unbounded we cannot conclude that f is not Lipschitz.

We return now to the initial value problem $y'=f(x,y)$, subject to $y(a)=\alpha$, and state a theorem which gives conditions sufficient to guarantee a unique, continuous solution in an interval containing $x=a$.

Theorem 2.1.1 Let $f(x,y)$ be continuous with respect to x and Lipschitz with respect to y in the rectangle

$$R:\ |x - a| \leqslant h,\ |y - \alpha| \leqslant k, \qquad h,k>0.$$

If $M=\sup|f(x,y)|$ and $\tau=\min(h,{}^{k}/_{M})$ then there exists a unique continuous solution, $y(x)$, of the differential equation $y'=f(x,y)$, on the interval $|x - a| \leqslant \tau$, which satisfies $y(a)=\alpha$.

Proof. See, for example, Coddington and Levinson (1955).

We now examine how well conditioned the problem $y'=f(x,y)$, $y(a)=\alpha$, is by considering some simple examples. The equation $y'=\lambda y$, where λ is a real constant and $y(0)$ is known, has the solution $y(x)=y(0)e^{\lambda x}$. If $\lambda>0$ a small change in $y(0)$ will produce a large change in $y(x)$ and so the problem is absolutely ill-conditioned. However, since $\lambda>0$ the solution itself will be large and so the error produced by a small discrepancy in $y(0)$ will form only a relatively small part of the solution; the problem is relatively well conditioned. If $\lambda<0$ this problem is absolutely and relatively well conditioned. Now consider $y'=\lambda(y-{}^{1}/_{3})$ which has the general solution $y(x)=Ae^{\lambda x} + {}^{1}/_{3}$ and if $y(0)={}^{1}/_{3}$ then $A=0$. However, the problem of representing ${}^{1}/_{3}$ in terms of a decimal or binary number means that instead of using $y(0)={}^1/_3$ it is necessary to use an approximation. For example, let us take $y(0)=0.3333$ in which case $A=-{}^1/_3\times10^{-4}$ and the solution becomes $y(x)=-{}^1/_3\times10^{-4}\ e^{\lambda x}+{}^1/_3$. If $\lambda<0$ then the error produced in this way will not be significant but if $\lambda>0$ then the exponential term will soon dominate the required solution. In this last case the problem is both absolutely and relatively ill-conditioned. Such ill-conditioning is by no means unusual; it is unrelated to the method of solution and is inherent within the problem itself. If we are to have any chance of computing a reasonable approximation for the solution of a differential equation such a solution must depend uniquely and continuously on the differential equation and any initial conditions which are to be imposed. That is to say that the solution of any small perturbation of the original problem must be close to the solution of the original problem. If this is so then the problem is said to be well posed. The existence of a Lipschitz constant, L, for f is sufficient to guarantee that if y_1 and y_2 are 2 solutions of $y'=f$, subject to $y_1(a)=\alpha$ and $y_2(a)=\alpha+\delta$, then $|y_1(x)-y_2(x)| \leqslant \delta\exp(L|x-a|)$.

2.2 Simple Difference Methods

The basis of many simple numerical techniques for solving the differential equation

$$y'=f(x,y), \quad y(a)=\alpha, \quad a\leqslant x\leqslant b, \qquad (2.2.1)$$

is to find some means of expressing the solution at x+h, i.e. $y(x+h)$, in terms of $y(x)$. In this way starting with the initial value $y(a)=\alpha$ an approximate solution can be generated at the discrete set of points $x=a+h, a+2h, \ldots$. For example, consider the Taylor series expansion of the solution about a, for then

$$y(a+h) = y(a) + hy'(a) + T(h), \qquad (2.2.2)$$

where the remainder term is given by $T(h)=\frac{h^2}{2}\,y''(\xi)$, $a<\xi<a+h$, and is frequently referred to as the **local truncation error.** If h is sufficiently small then we may approximate y(a+h) by the first two terms which, since y'=f(x,y), gives

$$y(a+h) \simeq y(a) + hf(a,y(a)). \qquad (2.2.3)$$

Therefore, if y(a) is known and a step size h is provided we can estimate y(a+h). This procedure is then repeated to obtain y(a+2h) from y(a+h) and so on. There is no real necessity to keep h fixed; in fact we shall see that it is advantageous to be able to adjust the step size h easily when necessary.

Example 2.2.1 Find an approximate solution of y'=y, y(0)=1.

Solution. Taking a step size h=0.1 gives:-

$$y(0.1) \simeq y(0) + 0.1\ f(0,y(0)) = y(0) + 0.1\ y(0) \simeq 1.1,$$
$$y(0.2) \simeq y(0.1)+0.1f(0.1,y(0.1)) \simeq 1.1 + 0.1\times 1.1 = 1.21,$$
$$y(0.3) \simeq y(0.2)+0.1\ f(0.2,y(0.2)) \simeq 1.331,$$
$$y(0.4) \simeq y(0.3)+0.1\ f(0.3,y(0.3)) \simeq 1.4641,$$
$$y(0.5) \simeq y(0.4)+0.1\ f(0.4,y(0.4)) \simeq 1.61051,$$

and so on. These values can be compared with the true solution $y(x)=e^x$ at the points x=0.1,0.2,0.3,0.4 and 0.5 as in Figure 2.2.1.

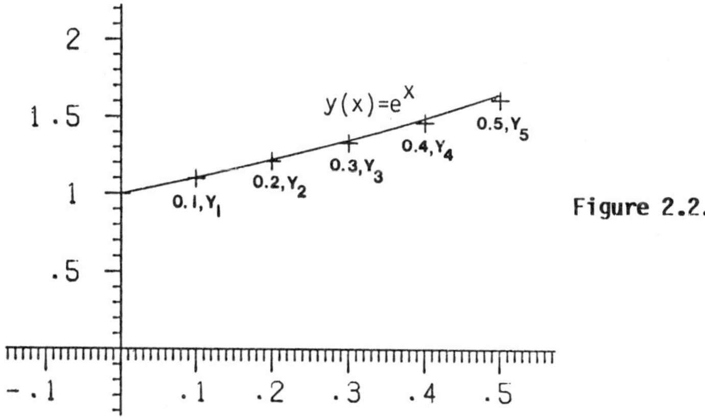

Figure 2.2.1

As $h \to 0$ the computed solution should tend to the true solution but this may not be the case. In Example 2.2.1, when y(0.1) was found it differed from the correct solution at x=0.1 since only the first two terms in the Taylor series were used to obtain equation (2.2.3). This error will produce subsequent errors in "y(0.2)", "y(0.3)" and so on. Whether this error eventually grows to such an extent that the computed solution bears no resemblance to the correct solution needs to be examined in detail.

Consider the differential equation $y'=f(x,y)$, subject to the initial condition $y(a)=\alpha$, for which we compute approximate values for y(a+h), y(a+2h), ..., y(a+nh), ... which will be denoted by $Y_1, Y_2, \ldots, Y_n, \ldots$. From equation (2.2.2) if h is sufficiently small then the difference solution, Y_n, may be generated by using the recurrence relation

$$Y_{n+1} = Y_n + hf(x_n, Y_n), \quad Y_0 = y(a) = \alpha, \; x_n = a+nh. \qquad (2.2.4)$$

This is called **Euler's Method.**

Exercise 2.2.1 Use Euler's method to find approximate solutions of

$$\text{(i)} \quad y'=xy^2, \quad \text{(ii)} \quad y'=\sqrt{x}\,\sin y,$$

subject to y(0)=1, in each case taking h=0.1, 0.05 and 0.025. Compare the computed approximations for y(1) for each equation.

We have seen that, at a fixed point, the discrepancy between the solution produced by equation (2.2.3) and the solution of the differential equation is determined by the local truncation error. We must now consider how such an error influences later values. Let us assume that an approximate solution has been computed beginning at x=a with a step size h. For a fixed value $x=x_n=a+nh$ the difference between y(x) and the corresponding discrete approximation is called the **global discretization error.** As $h \to 0$ increasingly many steps will be needed to reach x from any a, i.e. $n \to \infty$, and so a sufficient condition for the difference solution to converge to the solution of the differential equation is that the global error should tend to zero as $h \to 0$, $n \to \infty$ with $nh=(x-a)$.

Example 2.2.2 If Euler's method, (2.2.4), is applied to the equation $y'=\lambda y$, $y(0)=1$, where λ is a real constant, then the difference solution converges to the solution of the differential equation as the step size, h, tends to zero.

Solution. For the above differential equation Euler's method becomes

$$Y_{n+1} = Y_n + h\lambda Y_n = (1 + h\lambda)Y_n, \qquad (2.2.5)$$

which has the general solution $Y_n= A(1 + h\lambda)^n$, where A is an arbitrary constant. If A=1 then $Y_0=y(0)=1$ and so $Y_n=(1+h\lambda)^n$. For a fixed value of $x=x_n=nh$ the solution of the differential equation is $y(x)=e^{\lambda x}$ for which the corresponding difference approximation is

$$Y_n=(1 + {}^{x\lambda}/_{n})^n,$$

which converges to $y(x_n)=\exp(\lambda x_n)$ as $n \to \infty$. Notice that if $\lambda<0$ then the solution of the differential equation tends to zero. However, if h were chosen so that $1+h\lambda<-1$ the corresponding difference solution would grow increasingly large, successive terms having opposite sign and so the difference solution would bear no resemblance to the solution of the differential equation. This phenomenon is called **partial instability** and will be examined in Section 2.6. For this example it is necessary to select the step size so that $h<-{}^2/_{\lambda}$ in order to ensure that $Y_n \to 0$ as $n \to \infty$. If $\lambda>0$ the recurrence relation is relatively stable.

For the general equation $y'=f(x,y)$ it may be necessary to use a sequence of decreasing values for h and then examine the behaviour of the computed solution for signs of convergence.

Example 2.2.3 Find an approximate solution of $y'=y+x$, $y(0)=1$.

Solution. We use Euler's method as given by equation (2.2.4) and take $h=0.2$ and 0.1.

Table 2.2.1. Results of applying Euler's Method to $y'=y+x$, $y(0)=1$.

x_n	$h=0.1$ Y_n	$h=0.2$ Y_n	$h=0.1$ global error	$h=0.2$ global error
0.0	1.00000	1.00000	0.00000	0.00000
0.1	1.10000			
0.2	1.22000	1.20000	0.02281	0.04281
0.3	1.36200			
0.4	1.52820	1.48000	0.05544	0.10365
0.5	1.72102			
0.6	1.94312	1.85600	0.10112	0.18824
0.7	2.19743			
0.8	2.48718	2.34720	0.16390	0.30388
0.9	2.81590			
1.0	3.18748	2.97664	0.24908	0.45992

The general solution of this differential equation is $y(x)=-x-1+Ae^x$ for which the choice $A=2$ gives $y(0)=1$. Therefore, $y(1)=-2+2e$ or 3.4365637. Whilst the choice $h=0.2$ gives rather a poor result, taking $h=0.1$ gives slightly better agreement. If h were taken to be 0.01 then the computed approximation for $y(1)=y(100\times h)$ is Y_{100} which can be determined from equation (2.2.4) to be 3.4096276. Notice that the global error is proportional to h, i.e. reducing h by a factor of 2 halves the global error, despite the local truncation error being proportional to h^2.

Exercise 2.2.2 The initial value problem $y'=2ax+b$, $y(0)=0$, has the solution $y(x)=ax^2+bx$. Show that if Euler's method is applied to this problem then $Y_n=ax_n^2+bx_n-ahx_n$, where $x_n=nh$, and so $y(x_n)-Y_n=hax_n$ which tends to zero as $h\to 0$.

An alternative way of obtaining numerical methods is to integrate the differential equation over the interval $(x,x+h)$ and attempt to replace $\int f(x,y)dx$ by a numerical approximation, i.e.

$$y(x+h) - y(x) = \int_x^{x+h} f(x,y(x))dx = \int_x^{x+h} y'(x)dx. \qquad (2.2.6)$$

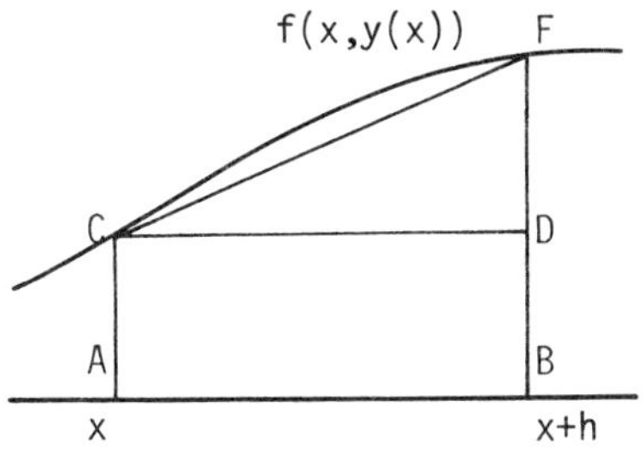

Figure 2.2.2

Approximating $\int f(x,y(x))dx$ by $hf(x,y(x))$, the area of the rectangle ABDC, gives Euler's method. A better scheme results if we replace the integral using the area of the trapezium ABFC, i.e.

$$\int_x^{x+h} f(x,y(x))dx = {}^h/_2\ [f(x+h,y(x+h)+f(x,y(x))] + T(h), \qquad (2.2.7)$$

where $T(h) = {}^{h^3}/_{12}\ {}^{d^3y}/_{dx^3}(\xi)$ and $x \leqslant \xi \leqslant x+h$. (See Section 1.5.3.) Combining equation (2.2.6) and (2.2.7) produces

$$y(x+h)=y(x)+{}^h/_2[f(x+h,y(x+h))+f(x,y(x))]+T(h). \qquad (2.2.8)$$

If h is sufficiently small and we replace $y(x)$ by Y_n, at $x=x_n$, then we obtain the difference method

$$Y_{n+1}=Y_n + {}^h/_2\ [f(x_n,Y_n) + f(x_{n+1},Y_{n+1})] \qquad (2.2.9)$$

which is called the trapezium method. This method is more accurate than the simpler Euler method, but it suffers from one major drawback. Examining equation (2.2.9) it is clear that Y_{n+1} is only given implicitly, and it will be necessary to solve an equation to find it. On the other hand, Euler's method gives Y_{n+1} explicitly.

Example 2.2.4 Find an approximate solution of $y' = xy^2$, $y(0)=1$, using the trapezium method (2.2.9).

Solution. Taking $h=0.1$ then Y_1 will be the computed approximation for $y(0.1)$, Y_2 for $y(0.2)$ and so on. From equation (2.2.9)

$$Y_1 = Y_o + {}^{0.1}/_2 \,[f(0,Y_o) + f(0.1,Y_1)] \qquad (2.2.10)$$

where $f(x,y)=xy^2$. Therefore,

$$Y_1 = 1 + 0.05[\,0 + 0.1Y_1^2]$$

or

$$0.005Y_1^2 - Y_1 + 1 = 0$$

which illustrates the problem of using an implicit method. It is necessary to solve a similar equation at each step. Solving this quadratic gives $Y_1=198.99494$ or $Y_1=1.005$ of which the former is clearly unrelated to the differential equation; the value $Y_1=1.005$ is used to find Y_2 and so on. In general, equations such as (2.2.10) would be solved by using a iterative method. For example, the iteration

$$(Y_1)^{r+1}=Y_o+0.05[f(x_o,Y_o)+f(x_1,(Y_1)^r)],$$

with $(Y_1)^o=Y_o$, converges linearly to Y_1 provided $|{}^h/_2 \, {}^{\partial f}/_{\partial y}|<1$. This condition can be satisfied by making the step size h sufficiently small.

Exercise 2.2.3 Use the trapezium method to find approximate solutions of

$$\text{(i) } y' = xy^2, \qquad \text{(ii) } y' = \sqrt{x}\,\sin y,$$

subject to $y(0)=1$, taking $h=0.1$, $h=0.05$ and $h=0.025$. Compare the computed approximations for $y(1)$.

Exercise 2.2.4 Show that when the trapezium method, (2.2.9), is applied to the test equation $y'=\lambda y$, where λ is a complex constant and $y(0)=1$, it produces a difference solution which converges to the correct solution as $h\to 0$. Show also that the recurrence relation associated with this method is stable if $|{}^{(2-\lambda h)}/_{(2+\lambda h)}|\leq 1$, i.e. the real part of λ is negative.

The difficulty in using the implicit trapezium rule offsets the advantage gained by the reduction in error achieved by this method. However, more accurate methods can be derived by including additional terms in the Taylor series expansion of y given by equation (2.2.2). Including one additional term gives

$$y(x+h) = y(x) + hy'(x) + {}^{h^2}/_2 y''(x) + T,$$

where the local truncation error, T, is proportional to h^3. However, $y'=f(x,y)$ and so

$$y'' = {}^{d}/_{dx}(y') = {}^{d}/_{dx}(f) = f_x + f_y\, {}^{dy}/_{dx} = f_x + f_y\, f,$$

where f_x and f_y denote the first partial derivatives of f with respect to x and y respectively. If T is sufficiently small and Y_n is the computed approximation for $y(x)=y(x_n)$ then

$$Y_{n+1} = Y_n + hf + {}^{h^2}/_2\, [f_x + f\, f_y], \qquad (2.2.11)$$

where f, f_x and f_y are evaluated at (x_n,Y_n). This gives Y_{n+1} explicitly in terms of Y_n and is as accurate as the trapezium method, however, to give this accuracy it is necessary to evaluate each of f, f_x and f_y at each step of the process. When f is at all complicated then its differentiation and evaluation can make the method (2.2.11) unsuitable. If we are willing to compute the higher order derivatives of f in the same way then it is possible to obtain methods of arbitrarily high accuracy.

2.3 General Linear Single Step Methods

All the methods developed so far have used only $y(x)$ to produce an approximation for $y(x+h)$, i.e. Y_{n+1} is given in terms of Y_n alone and, therefore, are called single step methods. Euler's method (2.2.4) and the trapezium method (2.2.9) are examples of the general linear single step method given by

$$Y_{n+1}+\alpha\, Y_n=h[\beta_0 f(x_{n+1},Y_{n+1})+\beta_1 f(x_n,Y_n)], \qquad (2.3.1)$$

where α_1, β_0 and β_1 are constants. The term linear refers to the fact that the method involves only a linear combination of terms involving f. The developed Taylor series method (2.2.10) is an example of a non-linear method. If $\beta_0=0$ then the method (2.3.1) gives Y_{n+1} explicitly otherwise it is given implicitly. For example Euler's method is explicit and has $\alpha_1=-1$, $\beta_0=0$, $\beta_1=1$, whereas the trapezium method is implicit and has $\alpha_1=-1$, $\beta_0 = \beta_1 = {}^1/_2$. Such methods are obtained by considering expressions of the form

$$y(x+h) + \alpha_1 y(x) = h[\beta_0 f(x+h,y(x+h))+\beta_1 f(x,y(x))] + T$$

at $x=x_n$. See for example equations (2.2.8) and (2.2.9). The object being to select those values of α_1, β_0 and β_1 which give the smallest value for the local truncation error T. Re-arranging the above expression for T and expanding in a Taylor series about x gives

$$\begin{aligned} T = \quad & y(x) + hy'(x) + h^2y''(x)/2 + 0(h^3) \\ & +\alpha_1 y(x) \\ & - \beta_0 hy'(x) - \beta_0 h^2 y''(x) - 0(h^3) \\ & - \beta_1 hy'(x), \end{aligned}$$

where $0(h^3)$ denotes those terms which involve powers of h in excess of 2. As T must tend to zero as $h\to 0$ it is necessary that $1+\alpha_1=0$, but by choosing

$$1 - \beta_1 - \beta_0 = 0, \qquad {}^1/_2 - \beta_0 = 0$$

T will be proportional to h^3. This gives the optimal coefficients $\alpha_1=-1$, $\beta_0=\beta_1={}^1/_2$, i.e. the trapezium method, together with

$$T = -\,{}^{h^3}/_{12}\;{}^{d^3y}/_{dx^3} + O(h^4). \qquad (2.3.2)$$

Notice that for Euler's method ($\alpha_1=-1$, $\beta_0=0$, $\beta_1=1$)

$$T = h^2\,{}^{y''(x)}/_2 + O(h^3). \qquad (2.3.3)$$

Comparing these expressions it is clear that if h is reduced to ${}^h/_2$ then the local error associated with the trapezium method is reduced by a factor of 8 while with Euler's method it is reduced by 4. For the highest accuracy it is clear that the local truncation error should be made proportional to the highest power of h possible.

2.4 General Single Step Methods

We now extend the idea of linear single step methods to more general techniques. Consider the general single step method given by

$$Y_{n+1} = Y_n + h\phi(x_n,Y_n,h),\ x_n=a+nh,\ Y_0=\alpha, \quad (2.4.1)$$

where Y_n is a computed approximation for $y(x_n)$. For example, Euler's method has $\phi(x,y,h)=f(x,y)$. Equation (2.4.1) can be written in the form

$$\frac{Y_{n+1} - Y_n}{h} = \phi(x_n,Y_n,h)$$

which is a discrete approximation for the differential equation $y'=f(x,y)$ since

$$\underset{h\to 0}{\text{Limit}}\ \frac{y(x+h) - y(x)}{h} = y' = f.$$

The term

$$E(h,x) = \frac{y(x+h) - y(x)}{h} - \phi(x,y,h) \qquad (2.4.2)$$

is a measure of the **discretization error** between the difference equation (2.4.1) and the differential equation $y'=f(x,y)$. If the discretization error tends to zero as $h\to 0$, for $a\leq x\leq b$, then

$$0 = \lim_{h\to 0} E(h,x) = \lim_{h\to 0} \left(\frac{y(x+h) - y(x)}{h}\right) - \phi(x,y(x),0)$$

$$= {}^{dy}/_{dx} - \phi(x,y,0)$$

and so the function ϕ must satisfy $\phi(x,y(x),0) = f(x,y)$. Re-arranging equation (2.4.2) gives

$$y(x+h) = y(x) + h\phi(x,y,h) + hE(h,x)$$

from which the relationship between the local truncation error and the discretization error is clear.

Definition 2.4.1 Let M be a positive constant and k the smallest positive integer such that $\sup|E(h,x)| \leq Mh^k$, for all sufficiently small h, then the difference method (2.4.1) is said to be **consistent** of order k with the differential equation $y'=f(x,y)$, $y(a)=\alpha$.

Exercise 2.4.1 Show that Euler's method is consistent of order 1 and that the trapezium method is consistent of order 2.

Having considered how accurately the difference method (2.4.1) approximates the differential equation (consistency) we now turn to the question of how well the solution of the difference equation approximates the solution of the differential equation (convergence). In the case of Euler's method, as applied to the test problem $y'=\lambda y$, this has already been shown. A similar result is true for the trapezium method given by equation (2.2.9). We now consider the convergence of the general single step method (2.4.1).

Theorem 2.4.1 Let the general single step difference method

$$Y_{n+1} = Y_n + h\phi(x_n,Y_n,h), \; Y_0=\alpha, \; x_n=a+nh, \; n=0,1,2,.. \qquad (2.4.3)$$

be consistent, of order k, with the differential equation

$$dy/dx = f(x,y), \quad y(a) = \alpha, \qquad a\leqslant x\leqslant b,$$

that is there exists a constant M such that $\sup|E(h,x)|\leqslant Mh^k$. In addition let ϕ be Lipschitz with respect to its second argument, then for any (x,y) and (x,z) in $[a,b]\times[\alpha-\tau,\alpha+\tau]$, $\tau > 0$, there exists a constant $L>0$ such that

$$|\phi(x,y,h) - \phi(x,z,h)| \leqslant L\,|y - z|. \qquad (2.4.4)$$

Under these conditions the global error satisfies

$$|y(x_n)-Y_n| \leqslant M/L\; h^k\,(\exp((x_n - a)L) - 1), \qquad (2.4.5)$$

for $a\leqslant x_n\leqslant b$ and sufficiently small h and the difference solution converges to the solution of the differential equation as $h\to 0$.

Proof. The difference method (2.4.3) is obtained by considering expressions of the form (2.4.2),

$$y(x_{n+1}) = y(x_n) + h\phi(x_n,y(x_n),\, h) + hE(h,x_n). \qquad (2.4.6)$$

Subtracting (2.4.3) from (2.4.6) and taking moduli gives

$$|y(x_{n+1})-Y_{n+1}| \leqslant |y(x_n)-Y_n| + h|\phi(x_n,y(x_n),h)-\phi(x_n,Y_n,h)|+h|E(h,x_n)|. \qquad (2.4.7)$$

Writing $e_n=y(x_n)-Y_n$ then

$$|e_{n+1}| \leqslant (1+hL)|e_n|+h|E|, \; e_0=0.$$

Now compare $|e_n|$ with w_n where

$$w_{n+1} = (1 + hL)w_n + Mh^{k+1}, \quad w_0=0. \qquad (2.4.8)$$

Clearly, since $|e_0|=0$, we have that $|e_0|\leq w_0$. Now suppose that $|e_r|\leq w_r$ for all $r\leq n$ then

$$|e_{n+1}| \leq (1 + hL)|e_n| + h|E| \leq (1 + hL)w_n + Mh^{k+1} = w_{n+1},$$

since $|E|\leq Mh^k$, and so, by induction, $|e_n|\leq w_n$ for all n. Since $L>0$ the general solution of (2.4.8) is $w_n = A(1 + hL)^n - {}^M/_L\ h^k$. However, as $w_0=0$ the arbitrary constant, A, is given by $A={}^M/_L h^k$, therefore, $|e_n| \leq w_n = {}^M/_L\ h^k\ [(1 + hL)^n -1]$. Furthermore, since $hL>0$ we have that $1+hL<e^{hL}$ and so

$$|e_n| \leq {}^M/_L\ h^k\ [\ e^{nhL} - 1] = {}^M/_L\ h^k\ [\exp(L(x_n-a))-1]$$

which tends to zero as $h\to0$, provided $k>0$, which establishes convergence.

Theorem 2.4.1 does more than establish the convergence of suitable one step methods since the expression (2.4.5) gives a bound on the difference between the computed and correct solutions. In some cases it is possible to obtain an explicit expression for L and M and hence estimate the global error. However, in reality to compute such a bound depends on being able to evaluate the k+1th derivative of the solution and although this is possible the bound obtained in this way is usually a severe overestimate. In general if

$$|y(x_n) - Y_n| \leq M\ h^k \qquad (2.4.9)$$

then the method is said to be **convergent of order** k, but as it is consistent of order k we merely say that it is of **order** k.

Example 2.4.1 Show that Euler's method is convergent of order 1.

Solution. Euler's method is given by $Y_{n+1}=Y_n+hf(x_n,Y_n)$, therefore $\phi(x,y,h) = f(x,y)$. From Exercise 2.4.1 we have that this method is consistent of order 1. Furthermore, since we have assumed that $f(x,y)$ is Lipschitz with respect to y, in order to guarantee a unique solution of the differential equation, the global error satisfies equation (2.4.5) with $k=1$. Therefore, Euler's method is convergent of order 1.

When dealing with recurrence relations of the form

$$Y_{n+1} = Y_n + h\phi(x_n,Y_n,h)$$

the problem of the accumulating rounding errors introduced at each step needs to be considered. By reducing h to control the discretization error we necessarily increase the number of steps required and may, therefore, increase the overall rounding error to such an extent that no benefit accrues. Let us suppose that at each step we in fact compute

$$Z_{n+1} = Z_n + h\phi(x_n,Z_n,h) + r_n,$$

where r_n is due to any possible rounding error at the nth step. Analogous with equation (2.4.7) we have that

$$|y(x_{n+1})-Z_{n+1}| \leqslant |y(x_n)-Z_n|+h|\phi(x_n,y(x_n),h)-\phi(x_n,Z_n,h)| + h|E(h,x_n)| + |r_n|$$

and if $|r_n| \leqslant \varepsilon$, for all n, then an extra term $^{\varepsilon}/_{h}$ must be added to $^{M}/_{L}h^k$ in equation (2.4.5) as a correction. The actual accuracy will therefore depend upon whether $^{\varepsilon}/_{h}$ is appreciable when compared with Mh^k as h tends to zero. In most cases $^{\varepsilon}/_{h}$ is negligible, but if not, then the term $Mh^k+^{\varepsilon}/_{h}$ has a minimum when $h^{k+1}=^{\varepsilon}/_{Mk}$.

2.5 Runge-Kutta Methods

In Section 2.3 it was shown that by introducing 3 parameters into a linear single step method the truncation error could be made proportional to h^3 giving a method of order 2. In general with r parameters it is possible to derive methods which are of order at most r-1. In this and the next section we shall give details of such methods. Let Y_n be an approximation for $y(x_n)$, $x_n=a+nh$, and then define the difference method

$$Y_{n+1} = Y_n + [\alpha_1 k_1 + \alpha_2 k_2 + \ \dots \ + \alpha_p k_p], \qquad (2.5.1)$$

where $k_1=hf(x_n,Y_n)$ and $k_j=hf(x_n+\nu_j h,Y_n+\mu_j k_{j-1})$ for $j=2,3,\dots,p$. This gives a total of p α's, p-1 ν's and p-1 μ's; 3p-2 parameters in all. The non-linear way in which the parameters ν_j and μ_j, $j=2,..,p$, occur complicates the situation, but if p is not too large then it is relatively straightforward to find suitable values.

p=1: This gives $Y_{n+1} = Y_n + \alpha_1 k_1$ and $k_1=hf(x_n,Y_n)$ or

$$Y_{n+1} = Y_n + h\alpha_1 f(x_n,Y_n)$$

which for $\alpha_1=1$ reduces to Euler's method.

p=2: This gives $Y_{n+1}=Y_n+[\alpha_1 k_1 + \alpha_2 k_2]$

$$k_1=hf(x_n,Y_n), \quad k_2=hf(x_n+\nu_2 h,Y_n+\mu_2 k_1).$$

To find suitable values for α_1, α_2, ν_2 and μ_2 recall that Y_n is an approximation for $y(x_n)$ and consider

$$y(x+h)=y(x)+h\alpha_1 f(x,y(x))+h\alpha_2 f(x+\nu_2 h,y(x)+\mu_2 hf(x,y(x)))] + T.$$

Expanding $f(x+\nu_2 h,y(x)+\mu_2 hf(x,y(x)))$ in a Taylor series gives

$$y(x+h)=y(x)+\alpha_1 hf(x,y(x))+\alpha_2 hf(x,y(x))+\alpha_2\nu_2 h^2 f_x +\alpha_2\mu_2 h^2 f(x,y(x))f_y+T,$$

which can be compared with the Taylor series expansion

$$y(x+h) = y(x) + hy'(x) + {}^{h^2}/_2\, y''(x) + O(h^3).$$

Since $y'=f(x,y(x))$ the chain rule gives $y''=d^2y/dx^2=f_x+f\times f_y$ and so

$$\alpha_1 + \alpha_2 = 1,\ \alpha_2\nu_2 = \alpha_2\mu_2 = {}^1/_2,$$

together with an expression for T which is proportional to h^3. This gives 3 equations in 4 unknowns and so there will not be a unique solution. For example $\alpha_1=\alpha_2={}^1/_2$, $\nu_2=\mu_2=1$ gives

$$Y_{n+1}=Y_n+{}^h/_2\,[f(x_n,Y_n)+f(x_n+h,Y_n+hf(x_n,Y_n))), \qquad (2.5.2)$$

which is called the **Simple Runge-Kutta** or **Improved Tangent Method.** The choice $\alpha_1=0$, $\alpha_2=1$, $\nu_2=\mu_2={}^1/_2$ gives

$$Y_{n+1}=Y_n + hf(x_n+{}^h/_2,Y_n+{}^h/_2 f(x_n,Y_n)), \qquad (2.5.3)$$

which is called the **Euler-Cauchy Method.** The method (2.5.2) bears a striking similarity to the trapezium method but it is important to recall that the trapezium method is implicit whereas the above method is explicit by design. It is often possible to decide between different Runge-Kutta methods of the same order on more positive grounds. For example the coefficients might be chosen to minimize the error T which for this example is given by

$$T=h^3\{({}^1/_6-{}^{\nu_2}/_4)({}^\partial/_{\partial x}+f{}^\partial/_{\partial y})^2 f +f_y({}^\partial/_{\partial x}+f{}^\partial/_{\partial y})f\}. \qquad (2.5.4)$$

There is no choice of the parameters which makes the expression (2.5.4) vanish identically but $\nu_2={}^2/_3$ produces an optimal formula. This gives $\alpha_1={}^1/_4$, $\alpha_2={}^3/_4$ and $\mu_2={}^2/_3$.

The relationship between the extended Taylor series method and the Runge-Kutta techniques is clear. The latter is an attempt to take advantage of the Taylor series without explicitly evaluating the partial derivatives of f. Clearly as p increases then the derivation of these methods is similar but tedious and time consuming. Taking p=4 gives one of the most widely used Runge-Kutta methods. For this choice of p a similar derivation produces 10 equations in 12 unknowns and so we can select any two parameters arbitrarily. For example, the **Classical Runge-Kutta Method** is

$$Y_{n+1}=Y_n + {}^1/_6[k_1+2k_2+2k_3+k_5], \qquad (2.5.5)$$

where

$$k_1=hf(x_n,Y_n),\ k_2=hf(x_n+{}^h/_2,Y_n+{}^1/_2k_1),$$
$$k_3=hf(x_n+{}^h/_2,Y_n+{}^1/_2k_2),\ k_4=hf(x_n+h,Y_n+k_3).$$

This method is of order 4. Notice that if f is independent of y then this method reduces to Simpson's rule for the numerical integration of f between x and x+h.

The choice p=5 produces no improvement in accuracy and to get a 5th order method it is necessary to take p=6. To obtain a 6th order method it is necessary to use p=7 or 8 whilst for $p \geq 9$ we obtain Runge-Kutta methods of order at most p-2. Clearly as p increases the number of function evaluations per step also increases and to some extent offsets the improvement in accuracy. Runge-Kutta methods of such high orders are given by Jain (1979) who considers higher order formulae in detail.

Example 2.5.1 Find an approximate solution of $y'=xy^2$, $y(0)=1$, using the Simple Runge-Kutta Method.

Solution. The Simple Runge-Kutta Method is given by equation (2.5.2). Taking h=0.1 gives the following results.

At x=0, $k_1=hf(0,Y_0)=0$, $k_2=hf(0.1,Y_0+k_1) = 0.01$,

$$Y_1 = Y_0 + 0.5\times[k_1+k_2] = 1.005.$$

At $x=h=0.1$, $k_1=hf(0.1,Y_1)=0.010100$ $k_2=hf(0.2,Y_1+k_1)=0.02060857$,

$$Y_2 = Y_1 + 0.5\times[k_1+k_2] = 1.02035441.$$

At $x=0.2$, $k_1=hf(0.2,Y_2)=0.0208224625$, $k_2=hf(0.2,Y_2+k_1)=0.0325214785$,

$$Y_3 = Y_2 + 0.5\times[k_1+k_2] = 1.04702638$$

similarly $Y_4=1.08679464$ and $Y_5=1.14256824$. This last value is the computed approximation for $y(x_5)=y(5h)=y(5\times0.1)=y(0.5)=1.14285714$. Using the same method with twice as many intervals, i.e. $h=0.05$, the corresponding value for $y(0.5)=y(10\times0.05)$ is Y_{10}; a little calculation gives $Y_{10}=1.14280493$ which is a somewhat better.

Example 2.5.2 Show that the Simple Runge-Kutta Method is convergent of order 2.

Solution. The Simple Runge-Kutta Method is given by equation (2.5.2) which can be written $Y_{n+1}=Y_n+h\phi(x_n,Y_n,h)$, where

$$\phi(x,y,h)={}^1/_2\ [f(x,y) + f(x+h,y+hf(x,y))].$$

If $f(x,y)$ is Lipschitz then, for all x and any y and z in some closed region R, there exists a constant L_0 such that $|f(x,y)-f(x,z)|\leqslant L_0|y-z|$. Therefore, for sufficiently small h,

$$\begin{aligned}|f(x+h,y+hf(x,y))-f(x+h,z+hf(x,z))| &\leqslant L_0|y+hf(x,y)-z-hf(x,z)| \\ &\leqslant L_0|y - z| + hL_0|f(x,y)-f(x,z)| \\ &\leqslant L_0(1 + hL_0)|y - z|.\end{aligned}$$

Hence

$$|\phi(x,y,h)-\phi(x,z,h)| \leqslant {}^1/_2\ [L_0 + L_0(1+hL_0)]|y-z|$$

and so ϕ is Lipschitz with constant $L=L_0(1+{}^1/_2hL_0)$. From Theorem 2.4.1, since this method is consistent of order 2 and ϕ satisfies a suitable Lipschitz condition this method will produce a difference solution which converges to the solution of the differential equation as $h\to0$ and the method is order 2.

Exercise 2.5.1 Compare the Simple and Classical Runge-Kutta Methods by computing an approximate solution of

$$\text{(i) } y'=xy^2, \qquad \text{(ii) } y'=\sqrt{x}\sin y,$$

given $y(0)=1$. In each case take $h=0.1$, $h=0.05$ and $h=0.025$ and compare the computed solutions for values of x between 0 and 1.

One of the major drawbacks with Runge-Kutta methods is the difficulty of estimating the error made at each step though this can be overcome by various means. For example, when using the Classical Runge-Kutta method if we evaluate

$$k_5=hf(x+{}^{3h}/_4, Y_n+(5k_1+7k_2+13k_3-k_4)/_{32}) \qquad (2.5.6)$$

then the term $2h(-k_1+3k_2+3k_3+3k_4-8k_5)/_3$ gives an approximation for the local truncation error at the current step which has error $O(h^5)$. By adjusting h it is therefore possible to control the growth of any local error produced in this way. Notice that k_5 involves not only k_4 but also k_1, k_2 and k_3 which leads to the more general form of the Runge-Kutta method given by

$$k_i=hf(x_n+\nu_i h, Y_n + \sum_{j=1,..,i-1} \mu_{ji}k_j), \quad i=1,2,...,p, \quad (2.5.7)$$

together with (2.5.1). This in turn leads to methods such as the Runge-Kutta-Merson method

$$Y_{n+1}=Y_n + {}^1/_6\,(k_1 + 4k_4 + k_5), \qquad (2.5.8)$$

where

$$\begin{aligned}
k_1&=hf(x_n,Y_n),\\
k_2&=hf(x_n+{}^h/_3, Y_n + (k_1)/_3),\\
k_3&=hf(x_n+{}^h/_3, Y_n + (k_1+k_2)/_6),\\
k_4&=hf(x_n+{}^h/_2, Y_n + (k_1+3k_3)/_8),\\
k_5&=hf(x_n+h, Y_n + (k_1-3k_3+4k_4)/_2).
\end{aligned}$$

The method (2.5.8) is of order 4, as is the classical Runge-Kutta method, but the local truncation error at each step can be estimated using the expression

$$(2k_1-9k_3+8k_4-k_5)/30.$$

The Runge-Kutta-Fehlberg method uses a slightly different approach. For example, it can be shown that the method given by

$$Y_{n+1}=Y_n + (\tfrac{25}{216}k_1+\tfrac{1408}{2565}k_3+\tfrac{2197}{4104}k_4- \tfrac{1}{5}k_5), \qquad (2.5.9)$$

where

$$\begin{aligned}
&k_1=hf(x_n,Y_n),\\
&k_2=hf(x_n+\tfrac{1}{4}h,Y_n+\tfrac{1}{4}k_1),\\
&k_3=hf(x_n+\tfrac{3}{8}h,Y_n+\tfrac{3}{32}k_1+\tfrac{9}{32}k_2),\\
&k_4=hf(x_n+\tfrac{12}{13}h,Y_n+\tfrac{1932}{2197}k_1-\tfrac{7200}{2197}k_2+\tfrac{7296}{2197}k_3),\\
&k_5=hf(x_n+h,Y_n+\tfrac{439}{216}k_1-8k_2+\tfrac{3680}{513}k_3-\tfrac{845}{4104}k_4).
\end{aligned}$$

is of order 4. Since this method requires 5 function evaluations as against four for the classical Runge-Kutta it would appear to be less efficient; however if we compute

$$k_6=hf(x+\tfrac{1}{2}h,Y_n-\tfrac{8}{27}k_1+2k_2-\tfrac{3544}{2565}k_3+\tfrac{1859}{4104}k_4-\tfrac{11}{40}k_5)$$

then

$$Y^*_{n+1}=Y_n+(\tfrac{16}{135}k_1+\tfrac{6656}{12825}k_3+\tfrac{28561}{56430}k_4-\tfrac{9}{50}k_5 + \tfrac{2}{55}k_6) \qquad (2.5.10)$$

gives a 5th order method. Therefore,

(i) from equation (2.5.9):- $Y_{n+1} = y(x_{n+1}) + C_1h^5 + O(h^6)$,

(ii) from equation (2.5.10):- $Y^*_{n+1} = y(x_{n+1}) + C_2h^6 + O(h^7)$

and so the difference $Y_{n+1}-Y^*_{n+1} \simeq C_1h^5 + O(h^6)$ is an estimate for the error associated with the method given by (2.5.9); accordingly, 6 function evaluations give not only Y_{n+1} but also an estimate for the local truncation error. For both the Merson and Fehlberg variants values for each of the k_i are found at each step, the error involved is estimated and then it is decided if it is necessary to adjust the step length h.

Runge-Kutta methods can be extended to implicit formulae by extending the definition (2.5.7) to

$$k_i=hf(x_n+\nu_i h, Y_n + \sum_{j=1,..,p} \mu_{ji}k_j),\ i=1,2,..,p. \quad (2.5.11)$$

For example, for p=1 this gives the method

$$Y_{n+1} = Y_n + k_1, \quad (2.5.12)$$

where $k_1=hf(x + {}^1/_2h, Y_n + {}^1/_2k_1)$. Notice that k_1 is only given implicitly and it is necessary to solve for it at each step. (See Butcher (1964).)

2.6 Stability of Single Step Methods

If Euler's method is applied to the test equation $y'=\lambda y$ then the difference solution, Y_n, is given by $Y_{n+1}=(1+\lambda h)Y_n$. This is an unstable recurrence relation unless $|1+\lambda h|\leqslant 1$. (See Chapter 1, Section 1.) If $Re(\lambda)>0$ then even though the discrepancy between $y(x_n)$, the solution of the differential equation at $x=x_n$, and the computed approximation grows, the relative error remains bounded; consequently there should be few problems. If $Re(\lambda)<0$ then $y(x)$ tends to 0; but if h is selected such that $|1+\lambda h| > 1$ the difference solution, Y_n, will grow. Therefore, it is necessary that $|1+\lambda h| \leqslant 1$ in order that the solution of the difference equation behaves like the solution of the differential equation. To analyse stability in this way depends on the test equation itself; using $y'=\lambda y$ can be "justified" as follows. Near a point (x_0,y_0) we can write $y'=f(x,y)$ as

$$y'=f(x_0,y_0) + (y-y_0)f_y(x_0,y_0) + (x-x_0)f_x(x_0,y_0) + \ldots$$

and provided f does not change too rapidly this can be approximated by

$$y'=\lambda y + ax + b,$$

where $\lambda=f_y(x_0,y_0)$, $a=f_x(x_0,y_0)$, $b=f(x_0,y_0)-x_0f_x-y_0f_y$. This equation has solution $y=Ae^{\lambda x} + p(x)$, where $p(x)$ is a polynomial of degree at most 2; clearly this solution will be dominated by the exponential term and hence the test equation $y'=\lambda y$. In this way we can justify the test equation $y'=\lambda y$ but the argument is far from rigorous.

Definition 2.6.1 A difference method is said to be **A-stable** if when applied to the test problem $y'=\lambda y$, where $Re(\lambda) < 0$, all difference solutions tend to zero.

Definition 2.6.2 A method is said to be **partially stable** if when applied to the test problem $y'=\lambda y$, λ a complex constant, the corresponding recurrence relation is absolutely stable for some values of λ with real part less than zero. The values for which it is absolutely stable constitute the **region of stability.**

Example 2.6.1 Compute an approximate solution of $y'=-30y$, $y(0)=1$.

Solution. Applying Euler's method gives the results in Table 2.6.1. For stability it is necessary that $h<2/30 \simeq 0.0666$.

Table 2.6.1 Results of applying Euler's method to $y'=-30y$, $y(0)=1$.

x	h=0.1 $1+\lambda h=-2.00$	h=0.01 $1+\lambda h=0.70$	h=0.001 $1+\lambda h=0.97$	h=0.0001 $1+\lambda h=0.997$
0.0	1.000000	1.000000	1.000000	1.000000
0.1	-2.000000	0.028248	0.047553	0.049531
0.2	4.000000	0.000798	0.002612	0.002456
:	:	:	:	:
0.5	-32.000000	$0.1798\ 10^{-7}$	$0.2431\ 10^{-6}$	$0.2990\ 10^{-6}$
:	:	:	:	:
1.0	1024.000000	$0.3234\ 10^{-15}$	$0.5912\ 10^{-13}$	$0.8945\ 10^{-13}$
maximum error	1024.00	$-0.2154\ 10^{-1}$	$-0.2235\ 10^{-2}$	$-0.2240\ 10^{-3}$

Example 2.6.2 Show that Euler's method is not A-stable but that it is partially stable.

Solution. The test problem $y'=\lambda y$ produces the recurrence relation $Y_{n+1}=(1+\lambda h)Y_n$ for which the region of stability is shown in Figure 2.6.1. Since the region of stability does not constitute the entire half plane $\mathrm{Re}(\lambda h)\leqslant 0$ this method is not A-stable.

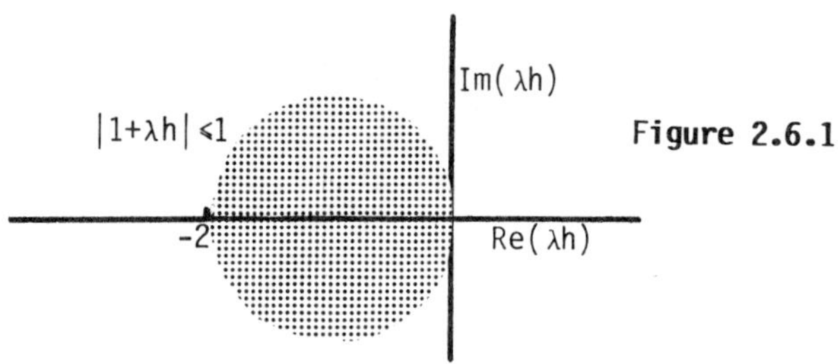

Figure 2.6.1

Example 2.6.3 Show that the trapezium method is A-stable.

Solution. From example 2.2.3 the region of stability of the trapezium method is the entire half plane $\mathrm{Re}(\lambda h)<0$ and so this method is A-stable.

Exercise 2.6.1 The Backward Euler method, $Y_{n+1}=Y_n+hf(x_{n+1},Y_{n+1})$, is an implicit two step method which is consistent of order 1. Find the region of stability for this method and confirm that it is A-stable. Show that the region of stability extends to that part of the half plane $\mathrm{Re}(\lambda h)>0$ which satisfies $|1-\lambda h|\geqslant 1$ and that for values of λ which satisfy this condition the solution of the differential equation is increasing but the computed difference solution, Y_n, tends to zero. This is called superstability.

Exercise 2.6.2 Show that the region of stability of the extended Taylor series method (2.2.11) and the simple Runge-Kutta method (2.5.2) is given by $|1+\lambda h+(\lambda h)^2/2|\leqslant 1$. (If λ is real: $-2\leqslant\lambda h\leqslant 0$.)

Exercise 2.6.3 Show that the region of stability for the Classical Runge-Kutta method (2.5.5), when applied to $y'=\lambda y$, is given by $|1+\lambda h+(\lambda h)^2/2+(\lambda h)^3/6+(\lambda h)^4/24|\leqslant 1$. (If λ is real: $-2.79<\lambda h\leqslant 0$.)

2.7 General Linear Multi-step Methods

In Section 2.3 it was shown that the best linear single step method was of order 2 and to achieve methods of higher order it was necessary to resort to non-linear methods such as Runge-Kutta schemes. We now consider formulae which are as accurate as the Runge-Kutta methods but which involve the function f linearly.

Example 2.7.1 Integrating $y' = f(x,y)$ between x-h and x+h gives

$$y(x+h) - y(x-h) = \int_{x-h}^{x+h} f(t,y(t))dt.$$

By approximating the integral in the simplest way find an explicit difference method which has local truncation error $O(h^3)$.

Solution. The simplest approximation for the above integral is given by the area of the rectangle ABCD in Figure 2.7.1, i.e.

$$y(x+h) - y(x-h) = 2hf(x,y(x)) + T(h). \qquad (2.7.1)$$

Expanding $y(x \pm h)$ in terms of a Taylor series about x gives that the local truncation error is $T(h)=h^3/3 \; d^3y/dx^3$ as required.

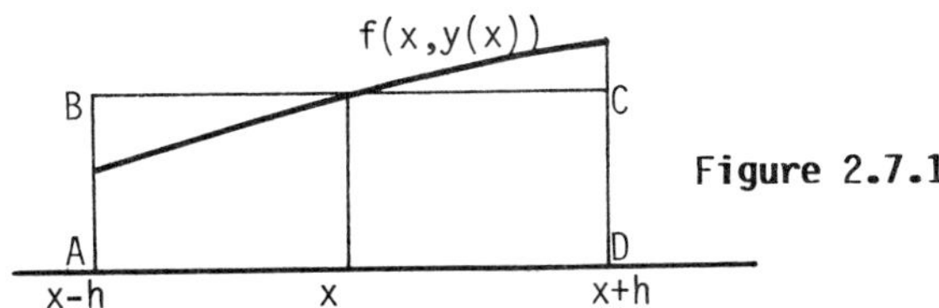

Figure 2.7.1

The local truncation error $T(h) \to 0$ as $h \to 0$, therefore, the explicit two step method

$$Y_{n+1} - Y_{n-1} = 2hf(x_n,Y_n) \qquad (2.7.2)$$

is consistent with the differential equation. We pay a penalty for retaining an explicit formulation because equation (2.7.2) is a second order recurrence relation which requires 2 starting values Y_0

and Y_1 in order to find Y_n, n>1, but the differential equation provides only one. Overlooking this point for the moment we consider the difference solution obtained when (2.7.2) is used to find an approximation for test equation $y'=\lambda y$, where λ is a real constant. Substituting this differential equation into (2.7.2) gives $Y_{n+1}-2h\lambda Y_n-Y_{n-1}=0$ which has the solution $Y_n=Ar_1^n+Br_2^n$ where A and B are arbitrary constants and r_1 and r_2 are the solutions of $p(r)=r^2-2h\lambda r-1=0$. We consider the cases $\lambda>0$ and $\lambda<0$ separately.

(i) $\lambda<0$,

$p(-1) = 2h\lambda < 0$,

$p(0) = -1 < 0$,

$p(1) = -2h\lambda > 0$,

Figure 2.7.2

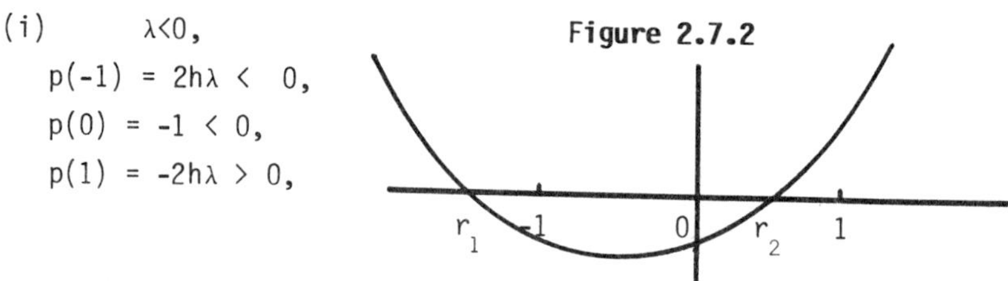

For all values of h, $r_2^n \to 0$ and so $Y_n \simeq Ar_1^n$ which oscillates with increasing amplitude whereas the true solution $y = y(0)e^{\lambda x}$ tends to zero. Therefore, for $\lambda<0$ the method is of no practical use.

(ii) $\lambda>0$,

$p(-1) = 2h\lambda > 0$,

$p(0) = -1 < 0$,

$p(1) = -2h\lambda < 0$.

Figure 2.7.3

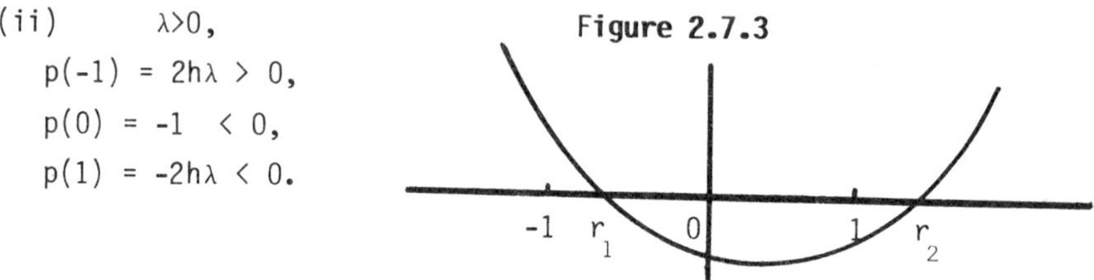

For all values of h $|r_1^n| \to 0$ and so $Y_n \simeq Br_2^n \to \infty$ but $y(x) \to \infty$ as well, in fact $r_2=1+\lambda h+O(h^2)$ and so $Y_n \simeq Br_2^n \simeq B(1+\lambda h+O(h^2))^n \to B\exp(\lambda x_n)$ as required.

The danger in employing a multi-step scheme is clear. Whereas the general solution of the differential equation has only one arbitrary constant, which may be obtained using the given initial condition, the general solution of a mth order linear recurrence relation is given by a linear combination of m independent solutions. Only one of these, called the principal solution, corresponds to the solution of the differential equation. The remaining m-1 solutions are called

parasitic and may or may not increase to such an extent that they dominate the required solution. Such instability should not be confused with the partial instability of single step methods, the problem here is the existence of increasing parasitic solutions. For Example 2.7.1 when $\lambda<0$ the solution required corresponds to r^n but r^n is the parasitic solution and in this case will soon dominate the required solution. When $\lambda>0$ r^n is the required solution and the parasitic solution r^n decays rapidly away.

Let us now reconsider the multi-step method (2.7.2) and note that it is a special case of the more general method

$$Y_{n+1}+\alpha_1 Y_n+\ldots+\alpha_m Y_{n-m+1} = h[\beta_0 f_{n+1}+\beta_1 f_n+\ldots+\beta_m f_{n-m+1}], \quad (2.7.3)$$

where f_p is used to denote $f(x_p,Y_p)$. This is the general linear multi-step method with step number m. If $\beta_0 = 0$ then the method gives Y_{n+1} explicitly otherwise it is given implicitly. When m=1 equation (2.7.3) reduces to the single step methods discussed in section 2.3. The difference method (2.7.2) has step number m=2 and $\alpha_1=0$, $\alpha_2=-1$, $\beta_0=0$, $\beta_1=2$, $\beta_2=0$.

Definition 2.7.1 The **local discretization error** associated with the m-step method (2.7.3) is defined to be

$$E(h,x)=\frac{y(x+h)+\alpha_1 y(x)+\alpha_2 y(x-h)+\ldots+\alpha_m y(x-(m-1)h)}{h} \quad (2.7.4)$$
$$- [\beta_0 y'(x+h)+\beta_1 y'(x)+\ldots+\beta_m y'(x-(m-1)h)].$$

Definition 2.7.2 Let k be the smallest positive integer such that $\sup|E(h,x)| \leqslant Mh^k$, $a\leqslant x\leqslant b$, M a constant, then the multi-step method (2.7.3) is said to be **consistent** of order k.

By expanding equation (2.7.4) in a Taylor series about x it is possible to express E as a power series in h and select the 2m+1 coefficients, $\alpha_1,\ldots,\alpha_m,\beta_0, \beta_1,\ldots,\beta_m$, to make E of order h^{2m-1}.

Example 2.7.2 Show that there exists a difference equation of the form

$$Y_{n+1}+\alpha_1 Y_n+\alpha_2 Y_{n-1} = h[\beta_0 f_{n+1}+\beta_1 f_n+\beta_2 f_{n-1}], \qquad (2.7.5)$$

where Y_n is a computed approximation for $y(x_n)$ and $f_n=y'(x_n)$, which is exact for all polynomials of degree less than 5.

Solution. Let $y_n=y(x_n)$ then the local discretization error of such a method is

$$\begin{aligned} E = {}^1/_h y_n & [1+\alpha_1+\alpha_2] \\ +y_n' & [1 \quad -\alpha_2-\beta_0-\beta_1-\beta_2] \\ +hy_n'' & [1 \quad +\alpha_2-2\beta_0 \quad -2\beta_2]/_2 \\ +h^2 y_n^{iii} & [1 \quad -\alpha_2-3\beta_0 \quad -3\beta_2]/_6 \\ +h^3 y_n^{iv} & [1 \quad +\alpha_2-4\beta_0 \quad -4\beta_2]/_{24} \\ +h^4 y_n^{v} & [1 \quad -\alpha_2-5\beta_0 \quad -5\beta_2]/_{120} + \ldots . \end{aligned}$$

As m=2 there are 2m+1=5 parameters to be selected and so we set the coefficients of ${}^1/_h, 1, h, h^2, h^3$ equal to zero which gives a system of 5 linear equations in 5 unknowns. Solving these equations gives $\alpha_1=0$, $\alpha_2=-1$, $\beta_0={}^1/_3$, $\beta_1={}^4/_3$, $\beta_2={}^1/_3$ or

$$Y_{n+1}-Y_{n-1}={}^h/_3[f_{n+1}+4f_n+f_{n-1}] + hE(h), \qquad (2.7.6)$$

where the local truncation error $T(h)=hE={}^{h^5}/_{180}\ {}^{d^5y}/_{dx^5}$ vanishes for polynomials of degree less than 5. This is one of a class of methods due to Milne. Notice that if f is independent of y then this method reduces to Simpson's rule for numerical integration.

Exercise 2.7.1 Find coefficients such that the difference method

$$Y_{n+1}+\alpha_1 Y_n+\alpha_2 Y_{n-1}+\alpha_3 Y_{n-2} = h[\beta_0 f_{n+1}+\beta_1 f_n+\beta_2 f_{n-1}+\beta_3 f_{n-2}], \qquad (2.7.7)$$

where $Y_n \simeq y(x_n)$ and $f_n=y'(x_n)$, is exact for all polynomials of degree less than 7.

The method given by equation (2.7.7) looks very attractive but in fact it suffers from catastrophic instability. Solving for the coefficients α_1, α_2, α_3, β_0, β_1, β_2 and β_3 produces the difference method

$$11Y_{n+1}+27Y_n-27Y_{n-1}-11Y_{n-2} = 3h[f_{n+1}+9f_n+9f_{n-1}+f_{n-2}]. \qquad (2.7.8)$$

For the test problem $y'=\lambda y$, $\lambda<0$, this reduces to

$$(11-3h\lambda)Y_{n+1} +(27-27h\lambda)Y_n + (-27-27h\lambda)Y_{n-1} + (-11-3h\lambda)Y_{n-2}=0.$$

The solutions of this recurrence relation are of the form $Y_n=Ar^n$, where r is a solution of the cubic equation

$$p(r)=(11-3h\lambda)r^3+(27-27h\lambda)r^2+(-27-27h\lambda)r+(-11-3h\lambda)=0.$$

Since $11-3h\lambda > 0$, $p(r)\to-\infty$ as $r\to-\infty$ but $p(-1)=32$ and so one solution of $p(r)=0$ is less than -1 for all values of λ. The presence of this parasitic solution, for all values of λ, will mean that the computed solution of the recurrence relation (2.7.8) will bear no resemblance to the solution of the differential equation which in this case is $y(x) = y(0)e^{\lambda x}$. This optimum method always suffers from this instability and is quite useless.

We now examine stability in a little more detail. With each difference equation of the form (2.7.3) we associate two m^{th} order polynomials

$$\rho(r) = r^m+\alpha_1 r^{m-1}+\alpha_2 r^{m-2} + \ldots +\alpha_m, \qquad (2.7.9)$$

and

$$\sigma(r) = \beta_0 r^m+\beta_1 r^{m-1} + \ldots\ldots + \beta_m. \qquad (2.7.10)$$

Therefore, the local discretization error, given by equation (2.7.4), is $E(h,x)=\rho(1)y(x)/h +(\rho'(1)-\sigma(1))y'(x) + O(h)$ and so for a consistent method we require that $\rho(1)=0$ and $\rho'(1)=\sigma(1)$. We consider now the stability, or otherwise, of such difference methods with respect to the test equation $y'=\lambda y$. The solutions of the resulting m^{th} order recurrence relation are of the form $Y_n=Ar^n$, where r is a zero of the polynomial function

$$p(r) = \rho(r) - \lambda h\sigma(r). \qquad (2.7.11)$$

In general p will be a polynomial of degree m and in order for the recurrence relation

$$Y_{n+1}(1-h\lambda\beta_0)+Y_n(\alpha_1-h\lambda\beta_1)+ \dots +Y_{n-m+1}(\alpha_m-h\lambda\beta_m)=0$$

to be absolutely stable it is necessary that all the zeros of p be within or at least on the unit circle. Certainly, if this is not the case then Y_n will grow exponentially even if the solution y(x) tends to zero. As the zeros of p depend upon h the determination of all the solutions of p(r) = 0 is very complicated, but since we require solutions in the limit as h tends to zero we consider instead the location of the zeros of ρ.

Definition 2.7.3 (Dahlquist). The linear multi-step method (2.7.3) is said to be **zero stable**, or simply **stable**, if all the zeros of $\rho(r)$ lie within or on the unit circle, and all the roots on the unit circle are simple. (This is also called D-stability.)

Clearly, stability defined in this way is a necessary condition for the zeros of p, given by equation (2.7.11), to lie within or on the unit circle.

Now that the concepts of consistency and stability have been defined we turn to the problem of establishing convergence. This problem has been considered in detail by Henrici (1968) and therefore we state the following results without proofs.

Theorem 2.7.1 A necessary condition for convergence of a linear multi-step method is that it is consistent of order at least 1.

Theorem 2.7.2 A necessary condition for convergence of a linear multi-step method is that it is stable.

This last result together with definition 2.7.3 is called the condition of stability and effectively considers the test problem $y'=0$. Notice that it is quite possible for a difference method to be stable according to definition 2.7.3 and yet produce an absolutely unstable recurrence relation. We illustrate the difference between the stability of the difference method and the recurrence relation produced by a specific differential equation using the following examples.

Example 2.7.3 The trapezium method $Y_{n+1}=Y_n+{}^h/_2(f_{n+1}+f_n)$ has $\rho(r)=r-1$, the only zero of which is at $r=1$, and is stable according to definition 2.7.3. Corresponding to equation (2.7.11) we have that $p(r)=r-1-{}^{h\lambda}/_2(r+1)$, which has its only zero at $r={}^{(2+h\lambda)}/_{(2-h\lambda)}$. This zero is within the unit circle if $Re(\lambda)<0$ and so the recurrence relation produced by the test equation $y'=\lambda y$ is stable in this range. When $Re(\lambda)>0$ the recurrence relation is relatively stable.

Example 2.7.4 For the multi-step method $Y_{n+1}-Y_{n-1} = 2hf(x_n,Y_n)$, $\rho(r)=r^2-1$ which has zeros at $r=1$ and $r=-1$ and so this method is stable according to definition 2.7.3. However, it has already been shown to produce an absolutely unstable recurrence relation for the test problem $y'=\lambda y$, $\lambda<0$. (See Example 2.7.1).

Example 2.7.5 The multi-step method (2.7.7) is not stable and the corresponding recurrence relation is also unstable for the test problem $y'=\lambda y$, $\lambda<0$. Since $\rho(r)=0$ has at least one root outside the unit circle, for all values of h, this method is said to be **strongly unstable.**

The concept of stability can be broadened to allow for problems with increasing solutions. In this case it may be quite acceptable for the errors involved to grow, but only if they form a relatively small part of the computed solution.

Definition 2.7.4 A linear multi-step method is said to be **relatively stable** with respect to the test problem $y'=\lambda y$ if all the zeros of the associated polynomial $\rho(r)$ are bounded by $e^{\lambda h}$.

We have already shown that the optimal 3 step method (2.7.7) is always unstable, in fact there is a restriction on the accuracy which can be obtained for any value of m. Henrici (1968) proves the following results.

Theorem 2.7.3 The order of a stable method whose step number m is odd is at most m+1.

Theorem 2.7.4 The order of a stable method whose step number m is even is at most m+2.

Example 2.7.6 The multi-step method (2.7.7) has step number 3 and has discretization error $O(h^6)$ but, by Theorem 2.7.3, its order must be less than 4 in order to be stable.

As stability is a necessary condition for convergence it would seem sensible to consider only difference schemes which satisfy such a criterion. We can do this by specifying the location of the zeros of $\rho(r)$ so that the condition of stability is satisfied. For consistency it is necessary that $\rho(1)=0$ and so ρ has a zero at r=1. This zero is frequently called the **principal zero.** Let us suppose that ρ has only a simple zero at r=1 and that all other zeros are at the origin, then for a m step method

$$\rho(r) = r^{m-1}(r - 1). \qquad (2.7.12)$$

This choice for ρ determines the so called Adams-Bashforth methods if we require explicit methods whilst if implicit methods are required we obtain the Adams-Moulton algorithms.

Example 2.7.7 The choice of $\rho(r)$ given by (2.7.12) produces the following difference methods.

Explicit. (Adams-Bashforth)

m=1.	$Y_{n+1}-Y_n=hf_n$, (Eulers Method),	(2.7.13)(i)
m=2.	$Y_{n+1}-Y_n=h[3f_n-f_{n-1}]/2$,	(2.7.14)(i)
m=3.	$Y_{n+1}-Y_n=h[23f_n-16f_{n-1}+5f_{n-2}]/12$,	(2.7.15)(i)

Implicit. (Adams-Moulton).

m=1.	$Y_{n+1}-Y_n= h[f_{n+1}+f_n]/2$,	(2.7.13)(ii)
m=2.	$Y_{n+1}-Y_n= h[5f_{n+1}+8f_n-f_{n-1}]/12$,	(2.7.14)(ii)
m=3.	$Y_{n+1}-Y_n= h[9f_{n+1}+19f_n-5f_{n-1}+f_{n-2}]/24$,	(2.7.15)(ii)

Exercise 2.7.2 Show that the leading terms in the discretization error of the above formulae are

(2.7.13) (i)	$h/2\ y''$	Order 1,
(2.7.13) (ii)	$-h^2/2\ d^3y/dx^3$	Order 2,
(2.7.14) (i)	$5h^2/12\ d^3y/dx^3$	Order 2,
(2.7.14) (ii)	$-h^3/24\ d^4y/dx^4$	Order 3,
(2.7.15) (i)	$-3h^3/8\ d^4y/dx^4$	Order 3,
(2.7.15) (ii)	$-19h^4/720\ d^5y/dx^5$	Order 4.

Notice that by Theorem 2.7.3 if m=1 then the order must be less than or equal to 2 and that if m=3 then the order must be less than or equal to 4. For m=2 the order cannot exceed 4 by Theorem 2.7.4.

The derivation of such multi-step methods is not very difficult mathematically but it can be a little messy. We illustrate, briefly, two alternative methods as a comparison with the Taylor series expansion described in Example 2.7.2.

Example 2.7.8 The difference operators Δ and ∇ are defined to be

$$\Delta Y_n = Y_{n+1} - Y_n, \qquad \nabla Y_n = Y_n - Y_{n-1}, \qquad (2.7.16)$$

for which it is possible to write

$$\Delta Y_{n+1} = h(1 + {}^1/_2 \Delta + {}^5/_{12}\ \Delta^2 + \ \ldots\ldots)Y_n', \qquad (2.7.17)(i)$$

and

$$\nabla Y_{n+1} = h(1 - {}^1/_2 \nabla - {}^1/_{12}\ \nabla^2 - \ \ldots\ldots)Y_{n+1}'. \qquad (2.7.17)(ii)$$

(For details of the Calculus of Differences see, for example, Burden, Faires and Reynolds, (1978).) The expression (2.7.17)(i) can be used to obtain the explicit Adams-Bashforth methods. For example, taking the first term gives $\Delta Y_{n+1} = hY_n'$, which is Euler's method $Y_{n+1} = Y_n + hY_n'$; using the first two terms gives the 2 step Adams-Bashforth method $Y_{n+1} = Y_n + h[3Y_n' - Y_{n-1}']/2$ as given by (2.7.14).

Exercise 2.7.3 Derive the Adams-Moulton methods from (2.7.17)(ii).

As a second alternative Henrici (1968) gives a algorithm for determining the polynomial σ given ρ. Since $\rho(1)=0$ the function ${}^{\rho(r)}/_{\log(r)}$ can be expanded in a power series about r=1, i.e.

$${}^{\rho(r)}/_{\log(r)} = c_0 + c_1(r-1) + .. + c_{m'}(r-1)^{m'} + \ldots$$

The choice $\sigma(r) = c_0 + c_1(r-1) + .. + c_{m'}(r-1)^{m'}$ gives a method which is of order at least m'+1. Note that the maximal order is m'+1 if m' is odd and m'+2 if m' is even. If m'=m+1 then the method obtained will be implicit, if m'=m then it will be explicit.

Example 2.7.9 Consider the two step method

$$Y_{n+1} - Y_{n-1} = h[\beta_0 f_{n+1} + \beta_1 f_n + \beta_2 f_{n-1}]$$

for which $\rho(r) = r^2 - 1$. Find the optimum implicit method.

Solution. From Theorem 2.7.4, since m=2, we should not expect a method of order in excess of 4. We have that

$$\rho(r)/\log(r) = 2 + 2(r-1) + {}^1/_3(r-1)^2 + O((r-1)^4)$$

and take

$$\sigma(r) = 2 + 2(r-1) + {}^1/_3(r-1)^2 = (1 + 4r + r^2)/3$$

which gives the fourth order Milne method (2.7.6).

Theorem 2.7.2 shows that stability is a necessary condition for convergence, we now give a result which guarantees that the Adams-Moulton algorithm is convergent provided it is consistent and the initial conditions required to start the recurrence relation are sufficiently accurate. If an m-step method is used then we need m values, $Y_0, Y_1, \ldots, Y_{m-1}$, before the recurrence relation can be used. These values need to be determined to at least the same order of accuracy as the difference method.

Theorem 2.7.5 Let the Adams-Moulton multi-step method

$$Y_{n+1}=Y_n + h[\beta_0 f_{n+1} + \ldots . + \beta_m f_{n-m+1}] \qquad (2.7.18)$$

be consistent, of order m+1, with the differential equation $y'=f(x,y)$, $y(a)=\alpha$, $a\leqslant x\leqslant b$. If the starting values $Y_0, Y_1, \ldots ., Y_{m-1}$ satisfy

$$|y(x_r) - Y_r| \leqslant M_r h^k, \quad r=0,1,\ldots,m-1,$$

for some $M_r > 0$ and sufficiently small h, then the global error satisfies

$$|y(x_n)-Y_n| \leqslant (\delta+{}^E/_{Lc})\exp[{}^{(x_n-a)Lc}/_{(1-hL|\beta_0|)}]-{}^E/_{Lc}, \qquad (2.7.19)$$

where $c=|\beta_0|+|\beta_1|+ \ldots +|\beta_m|$, L is a Lipschitz constant for f, $\delta=\max(M_r h^k)$, $E=\sup|E(h,x)|$ and $E(h,x)=h^{m+1}a_m y^{(m+2)}(\xi_n)$ is the local discretization error with a constant a_m particular to the method selected.

The inequality (2.7.19) provides a global error bound for the m-step Adams-Moulton algorithm; furthermore, as the local discretization error is $O(h^{m+1})$ and the starting values Y_r, $r=0,1,...,m-1$, are given to an accuracy of $\delta=O(h^k)$ the difference solution converges to the solution of the differential equation, as $h\to 0$, and is of order min(k,m+1). The starting values could be determined by high order Runge Kutta methods since such methods are single step and so do not require additional starting values in the same way.

The Adams-Bashforth and the Adams-Moulton methods were derived on the basis of $\rho(r)$ having only one zero with $|r|=1$ but it is possible to find alternative methods by specifying where the other zeros are to be located on the unit circle. For example, if ρ has zeros at $r=+1$ and $r=-1$ then we obtain the Nystrom methods:-

$$Y_{n+1}=Y_{n-1}+h[\beta_0{}^*f_{n+1}+\cdots\cdots+\beta_m{}^*f_{n-m+1}] \qquad (2.7.20)$$

which are explicit if $\beta_0{}^* = 0$, and implicit otherwise. We have already considered one such method in Example 2.7.1 where

$$Y_{n+1} = Y_{n-1} + 2hf(x_n,Y_n)$$

was shown to produce an absolutely unstable recurrence relation for the test equation $y'=\lambda y$, $\lambda<0$. Milne's method as given by equation (2.7.6) is also of the form (2.7.20). These methods are prone to instability since any non-zero value of h can shift a non-principal zero of ρ which lies on the unit circle to give a zero of p which lies outside it and so these methods are said to be **weakly stable.** The global error bound given by Theorem 2.7.5 does not apply to (2.7.20) but it is possible to prove the following general result.

Theorem 2.7.6 A stable and consistent linear mutli-step method converges.

The proof of Theorem 2.7.6 requires analytic result beyond the scope of this text; full details are given by Henrici (1968).

In practice implicit methods are to be preferred for in general they will have a smaller truncation error than an explicit method of the same order and they suffer less from practical instabilities. However, the application of an implicit method will mean that it is necessary to solve a non-linear equation for Y_{n+1} at each step.

Given the general implicit m-step method,

$$Y_{n+1}+\alpha_1 Y_n+..+\alpha_m Y_{n-m+1}=h[\beta_0 f_{n+1}+..+\beta_m f_{n-m+1}]$$

we may attempt to find the solution Y_{n+1} as suggested in Section 1.3 by defining a sequence of estimates $\{Y_{n+1}\}^r$, r=0,1,2,... which satisfy

$$\{Y_{n+1}\}^{r+1}+\alpha_1 Y_n+..+\alpha_m Y_{n-m+1} =$$

$$h[\beta_0 f(x_{n+1},\{Y_{n+1}\}^r)+..+\beta_m f_{n-m+1}] \qquad (2.7.21)$$

or equivalently

$$\{Y_{n+1}\}^{r+1} = h\beta_0 f(x_{n+1},\{Y_{n+1}\}^r) + C_n, \qquad (2.7.22)$$

where $C_n=h[\beta_1 f_n+...+\beta_m f_{n-m+1}]-\alpha_1 Y_n-...-\alpha_m Y_{n-m+1}$. The iteration (2.7.22) is preferred, for it shows that we need only compute C_n once for each n irrespective of the number of iterations that are required to find Y_{n+1}. If f is differentiable with respect to y then the iteration (2.7.22) will converge if

$$|h\beta_0 \,{}^{\partial f}/_{\partial y}| < 1, \qquad (2.7.23)$$

for all values of y close to Y_{n+1}. If f is not differentiable we merely replace $^{\partial f}/_{\partial y}$ by the Lipschitz constant which is associated with f. (For a simple illustration see Example 2.2.3.)

Example 2.7.10 Given the differential equation $y'=x^2+y^2$, subject to $y(0)=1$, determine conditions under which the iteration associated with the Trapezium method converges.

Solution. Using the recurrence relation $Y_{n+1}=Y_n+0.5h[f_{n+1}+f_n]$, the iteration corresponding to (2.7.22) converges if $|h\beta_0 f_y| < 1$. Since $f_y=2y$, $\beta_0=1/2$ we require that $h < 1/|y|$ near $y=Y_{n+1}$.

Exercise 2.7.4 Show that the iterations corresponding to the Adams-Moulton algorithms of order 2,3 and 4 converge if:-

$$|hL| < 2, \qquad \text{order } 2,$$
$$|hL| < 2.4, \qquad \text{order } 3,$$
$$|hL| < 2.67, \qquad \text{order } 4,$$

where L is the appropriate Lipschitz constant for f.

Finally, we examine the effect that rounding errors and the limitation of having to solve equations such as (2.7.21) iteratively have on convergence. For example, in Theorem 2.7.5 it was assumed that we could find Y_{n+1} exactly but as we have seen this may not be the case. However, we can amend this result by assuming that instead of equation (2.7.18) we have

$$Z_{n+1}=Z_n + h[\beta_0 f_{n+1}+\ldots+\beta_m f_{n-m+1}] + r_n + s_n, \qquad (2.7.24)$$

where r_n is any error due to rounding in the computation and s_n the error introduced because we cannot in general find Y_{n+1} exactly. If $|r_n| \leqslant r$ and $|s_n| \leqslant s$ then it is only necessary to add an additional term to equation (2.7.19) in which case $E=\sup|E(h,x)| + r + s$. If we have an estimate for E then it is possible to determine how large h should be if r and s are to be negligible.

Having considered the stability of linear multi-step methods we now return to see if they are A-stable. Unfortunately, the vast majority are not.

Figure 2.7.4 Stability Regions for Adams Methods of Order 2 and 3.

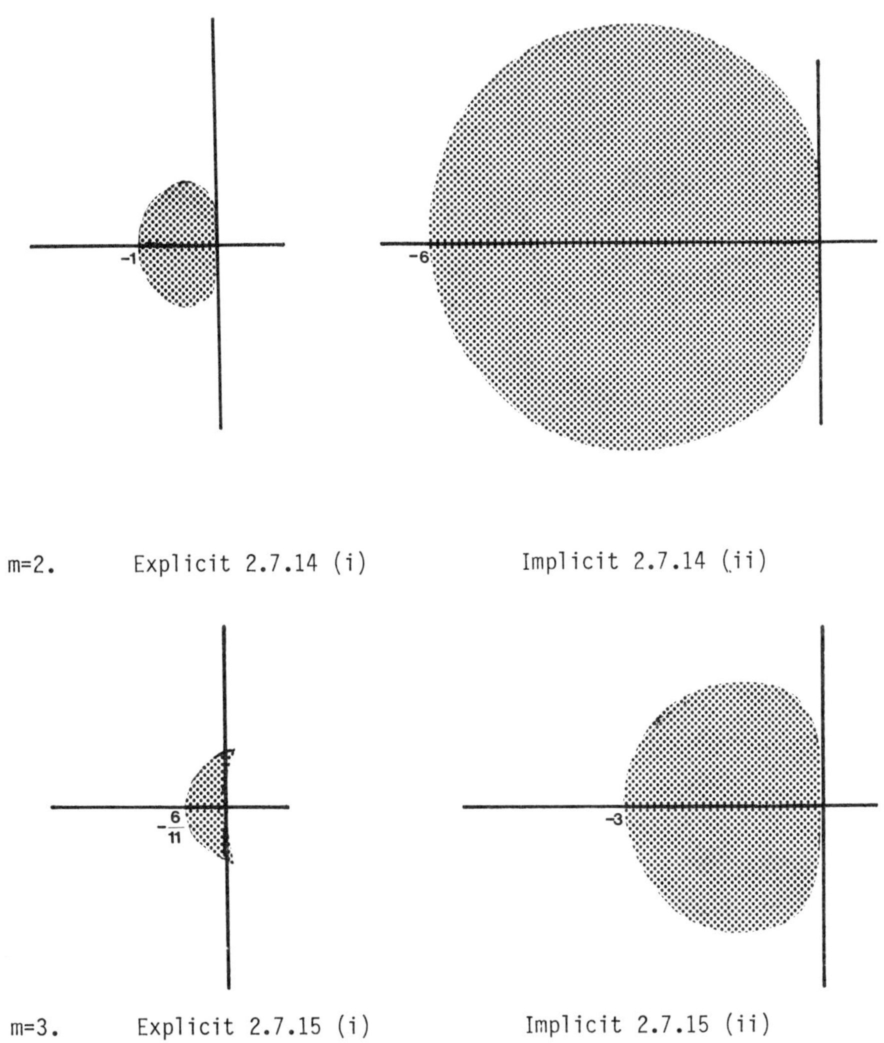

Theorem 2.7.7 The order of an A-stable linear multi-step method cannot exceed 2 and the method must be implicit.

Accordingly the only linear A-stable methods are the Trapezium method and the Backward Euler method $Y_{n+1}=Y_n+hf(x_{n+1},Y_{n+1})$. If we are considering a genuine multi-step method then we need to choose the various coefficients not only to give a stable and consistent method but one for which the region of stability is as large as possible.

Exercise 2.7.5 Show that the region of stability of the 2nd and 3rd order Adam's methods (2.7.14) and (2.7.15) are as given in Figure 2.7.4.

We have only analysed the stability of difference methods with respect to the test equation $y'=\lambda y$ but as argued in Section 2.6 given the equation $y'=f(x,y)$, provided f does not change too rapidly, we may take $\lambda \simeq \partial f/\partial y$ and by evaluating this derivative we can gauge whether or not we are within the region of stability of the relevant method.

Example 2.7.11 For the equation $y'=\sin\pi y$, $y(0)=1$, we have $\partial f/\partial y=\pi\cos\pi y$ which is approximately $-\pi$ near $y=1$. Therefore, for Euler's method to be stable we require that $|1-\pi h|\leqslant 1$ or $h\leqslant 2/\pi$. Likewise for the Adams-Bashforth and Adams-Moulton methods it is essential that the step size h be selected so that $-\pi h$ lies within the region of stability.

Exercise 2.7.6 Determine the maximum step size for which the Adams-Bashforth and Adams-Moulton methods of order 1,2 and 3, given in Example 2.7.7, are stable for the equation $y'=\sin\pi y$, $y(0)=1$.

2.8 Predictor-Corrector Methods

The speed with which the iteration (2.7.22) converges is largely dependent upon how close $\{Y_{n+1}\}^0$ is to Y_{n+1}. It seems sensible, therefore, to predict a reasonable value for Y_{n+1} using an explicit difference method and then to use this value as the initial estimate, $\{Y_{n+1}\}^0$, which is then corrected using an implicit formula by iterating to convergence. Such a procedure is called a Predictor-Corrector method and forms the basis of many of the most widely used methods for solving differential equations. The ultimate convergence, accuracy and stability of the method depends only upon the corrector formula which is used but there are several advantages in using a predictor which is of the same order as the corrector.

Example 2.8.1 Examples of Predictor-Corrector methods.

Predictor.	**Corrector.**	
$Y_{n+1}=Y_n + {}^h/_2(3f_n-f_{n-1})$	$Y_{n+1}=Y_n+h[f_{n+1}+f_n]/_2$,	(2.8.1)
$Y_{n+1}=Y_{n-3} + 4[2f_n-f_{n-1}+2f_{n-2}]/_3$,	$Y_{n+1}=Y_{n-1}+h[f_{n+1}+4f_n+f_{n-1}]/_3$,	(2.8.2)
	(Milne-Simpson method.)	
$Y_{n+1}=Y_n+h[55f_n-59f_{n-1}+37f_{n-2}-9f_{n-3}]/_{24}$,	$Y_{n+1}=Y_n+h[9f_{n+1}+19f_n-5f_{n-1}+f_{n-2}]/_{24}$,	(2.8.3)

If the predictor and corrector formula are of the same order then it is relatively straight forward to estimate the leading term in the local error. For example, for one of the most popular 4-th order Adam's predictor-corrector pairs (2.8.3) the local discretization errors are ${}^{251}/_{720}\ h^4 d^5y/_{dx^5}(\xi_n)$ and $-{}^{19}/_{720}\ h^4 d^5y/_{dx^5}(\zeta_n)$ where ξ_n and ζ_n lie between x_n and x_{n+1}, i.e.

$$y(x_n+h) = y(x_n)+h[55y'(x_n)-59y'(x_n-h)+37y'(x_n-2h)-9y'(x_n-3h)]/24$$
$$+ {}^{251}/_{720}\ h^5\ y^{v}(\xi_n)$$

and

$$y(x_n+h)=y(x_n)+h[9y'(x_n+h)+19y'(x_n)-5y'(x_n-h)+y'(x_n-2h)]/24$$
$$- {}^{19}/_{720}\ h^5\ y^{v}(\zeta_n)$$

the additional power of h in these equations being accounted for by the way the local discretization error has been defined. If we denote the predicted value by $Y_{n+1}{}^P$ and its corrected value by $Y_{n+1}{}^C$ then we have that

$$Y_{n+1}{}^P - Y_{n+1}{}^C \simeq {}^{251}/_{720} h^5 {}^{d^5y}/_{dx^5} + {}^{19}/_{720} h^5 {}^{d^5y}/_{dx^5} = {}^3/_8 h^5 {}^{d^5y}/_{dx^5},$$

provided h is sufficiently small for $\xi_n \simeq \zeta_n$. From this expression we can estimate ${}^{d^5y}/_{dx^5}$ and hence the local discretization error. A judicious choice of h can then be made to make the local discretization error suitably small. Note that the Milne-Simpson method, (2.8.2), is of the same order but that the corrector is only weakly stable.

Example 2.8.2 Find an approximate solution of $y'=y^2$, subject to $y(0)=1$ using the predictor-corrector pair (2.8.1).

Solution. First of all we note that the predictor of the pair is a two step method and so we need an additional starting value. Taking $h=0.1$ we find an approximation for $y(0.1) \simeq Y_1$ by using the Classical Runge Kutta method (2.5.5), this gives $Y_1=1.11111111$. To find $Y_2 \simeq y(0.2)$ we use the predictor to obtain

$$Y_2 = Y_1 + {}^{0.1}/_2(3Y_1^2 - Y_0^2) = 1.11111111+0.05(3(1.11111111)^2-1^2)$$
$$=1.2462962.$$

Now we correct this value using the implicit trapezium method written as a simple iteration, i.e.

$$\{Y_2\}^{r+1}=Y_1+0.05([\{Y_2\}^r]^2+Y_1{}^2),\ \{Y_2\}^o=1.246296.$$

This gives the sequence of values 1.2505022, 1.2510272, 1.2510929, 1.25110119, 1.25110222, 1.25110234, 1.25110236, 1.25110236. Therefore, the corrected value is given by $Y_2{}^C=1.25110236$ whilst $Y_2{}^P=1.2462962$. Further results are given in Table 2.8.1. The discretization errors, associated with the predictor-corrector pair (2.8.1), are given by $^5/_{12}\ h^2 d^3y/_{dx^3}$ and $-^1/_{12}\ h^2 d^3y/_{dx^3}$, and so

$$Y_2{}^P-Y_2{}^C = {}^6/_{12}h^3\ d^3y/_{dx^3} = 4.80606547\ 10^{-3},$$

therefore, the error in the corrected value is approximately $8.0101\ 10^{-4}$ compared with the actual error of $1.1023\ 10^{-3}$. Notice that this gives an estimate of the local error and is not a bound. We can use this technique to determine if the step size is sufficiently small. Let us suppose that we require the local error to be of order 10^{-5}. Since

$${}^6/_{12}\ h^3\ d^3y/_{dx^3} \simeq 4.806\ 10^{-3}$$

we have that

$$d^3y/_{dx^3} \simeq 9.612, \qquad (d^3y/_{dx^3}(0.2)=9.1449)$$

and to achieve the required order of error we take h so that

$${}^1/_{12}\ h^3\ 9.612 \leqslant 10^{-6},$$

therefore, we require $h\leqslant 0.00854632$. This would mean taking approximately 12 times as many steps but produces an approximation for $y(0.2)$ of 1.249976. We will now show that provided h is taken this small there is in fact no point in iterating the corrector formula to convergence.

Table 2.8.1 The approximate solution of $y'=y^2$, $y(0)=1$ using the Predictor-Corrector formulae (2.8.1).				
x	Predictor Only	Predictor Corrector	$P(EC)^1$ h=0.005	Exact Solution
0.2	1.24629620	1.25110236	1.25000463	1.25000000
0.3	1.41755607	1.43187910	1.42858185	1.42857143
0.4	1.64131313	1.67460871	1.66668874	1.66666666
0.5	1.94492618	2.01855204	2.00004756	2.00000000

We can make further use of the above estimate because ideally we would like to carry out only one iteration using the corrector formula. Let us assume that the predictor is of order k in which case the local discretization error at the nth step will be of the form $a_k h^k y^{(k+1)}$ whilst that of a k-th order predictor will be $b_k h^k y^{(k+1)}$. As in the previous example

$$Y_{n+1}{}^P - Y_{n+1}{}^C = (a_k - b_k) h^{k+1} y^{(k+1)}, \qquad (2.8.4)$$

from which we may estimate the k+1st derivative of y and hence $b_k h^k y^{(k+1)}$. If we were to use only one iteration of the corrector formula then it is necessary to ensure that $|\{Y_{n+1}\}^2 - \{Y_{n+1}\}^1|$ is sufficiently small. This can be done as follows:-

$$\begin{aligned} |\{Y_{n+1}\}^2 - \{Y_{n+1}\}^1| &= h\beta_0 |(f(x_{n+1}, \{Y_{n+1}\}^1) - f(x_{n+1}, \{Y_{n+1}\}^o)|, \\ &< h\ \beta_0\ L\ |\{Y_{n+1}\}^1 - \{Y_{n+1}\}^o|, \end{aligned}$$

where L is the Lipschitz constant for f with respect to y. As we intend to correct only once $\{Y_{n+1}\}^1 = Y_{n+1}{}^C$ and $\{Y_{n+1}\}^o = Y_{n+1}{}^P$ and so we require that

$$|\{Y_{n+1}\}^2 - \{Y_{n+1}\}^1| < \beta_0 L |a_k - b_k| h^{k+2} M_{k+1}, \qquad (2.8.5)$$

where M_{k+1} is an upper bound on the $k+1^{th}$ derivative of y. Therefore, the error due to using only one iteration of the corrector, given by (2.8.5), will be small compared with the local truncation error, $b_k h^k y^{(k+1)}$, if

$$h << |b_k / [L \beta_0 (a_k - b_k)]|.$$

Assuming that this is the case there is no advantage to be gained in applying the corrector formula more than once. To emphasise that only one iteration has been performed we say that such a method is of the form PEC which implies that only 1 application of the predictor, 1 evaluation of the function f, and 1 iteration of the corrector has been used. The general predictor corrector is of the form $P(EC)^m$ which implies that m applications of the corrector have been used. If the latter method is used, i.e. the corrector formula is iterated to convergence, and f is at all complicated it is frequently more efficient to reduce the step size h since this will mean that fewer iterations will be required and also will reduce the local error.

Exercise 2.8.1 Find an approximate solution of $y' = \sqrt{x}\,\sin y$, $y(0)=1$, using a suitable predictor-corrector method. Adjust the step size such that at each step the local error is of order 10^{-6}. How small need h be in order to make more than one iteration at each step unnecessary?

Using predictor and corrector formulae of the same order can be further exploited to extrapolate the computed values to give an improved estimate. Let us suppose that for some value $x=x_n$

$$y(x_n) \simeq Y^P + Ah^k, \quad y(x_n) \simeq Y^C + Bh^k$$

then

$$B(y(x_n) - Y^P) \simeq A(y(x_n) - Y^C)$$

or

$$y(x_n) \simeq (AY^C - BY^P)/(A-B). \qquad (2.8.6)$$

This leads to a class of so called Modified Predictor-Corrector methods. (See, for example, Jain (1979) for further details.)

Example 2.8.3 Applying the Predictor-Corrector (2.8.1) to $y'=y^2$, $y(0)=1$, with $h=0.1$, produces at $x=0.2$ $Y^P=1.2462962$ and $Y^C=1.25110236$. (See Example 2.8.2). From (2.8.6) we obtain the approximation $y(0.2) \simeq (5Y^C+Y^P)/_6=1.2503012$ as compared with the exact answer of 1.25.

The idea of extrapolation will be considered in more detail in the next section.

2.9 Extrapolation - Improvement of Solution

We have seen how it may be possible to control the local error by suitably adjusting the step size h. Now we examine an alternative based on the idea of deferred correction or Romberg extrapolation which was discussed briefly in Section 1.5.

The basis of this method is the fact that the difference between the true solution, $y(x)$, and a computed approximation at x calculated using a step size h, $Y(x,h)$, can be expressed as a function of h, i.e.

$$y(x) - Y(x,h) = g(h). \qquad (2.9.1)$$

If g is expressable in terms of simple functions of h it may be possible to eliminate such terms one by one. For example, if $g(h)=Ah^2+Bh^4+\ldots$ then we proceed as follows. We have that

$$y(x) - Y(x,h) = Ah^2 + Bh^4 + \ldots . \qquad (2.9.2)$$

$$y(x) - Y(x,{}^h/_2) = A({}^h/_2)^2 + B({}^h/_2)^4 + \ldots \qquad (2.9.3)$$

Multiplying equation (2.9.3) by 4 and subtracting the resulting equation from (2.9.2) gives

$$3y(x) - [4Y(x,{}^{h}/_{2})-Y(x,h)] = B'h^{4} + \ldots. \qquad (2.9.4)$$

where B' is a simple multiple of B. Therefore, the expression

$$[4Y(x,{}^{h}/_{2}) - Y(x,h)]/_{3} \qquad (2.9.5)$$

gives an approximation for $y(x)$ with error $O(h^4)$ rather that $O(h^2)$ as with $Y(x,h)$ and $Y(x,{}^{h}/_{2})$. In the same way we can eliminate the next term involving h^4 and then subsequent terms.

Example 2.9.1 Find an approximate value for the solution of $y'=y$, subject to $y(0)=1$, for $x=1$.

Solution. Using the simple trapezium method with step size $h=0.5$ gives $y(1) \simeq Y_2=2.77777\cdot$, for $h=0.25$ $y(1)=y(4\times 0.25) \simeq Y_4$ which is given by 2.7326114, and for $h=0.125$ $y(8\times 0.125) \simeq Y_8=2.72183189$. Notice, first of all, that ${}^{(Y(1,0.5)-Y(1,0.25))}/_{(Y(1,0.25)-Y(1,0.125))}$ is approximately 4 verifying that this method is of second order as was shown by Theorem 2.4.1. In fact it can be shown that the difference $y(1)-Y(1,h)$ satisfies (2.9.3) and using the expression (2.9.5) with $h=0.5$ gives 2.71755594 and with $h=0.25$ gives 2.7181238724. Likewise, eliminating the term of $O(h^4)$ gives 2.71828424 which has error $-2.4\ 10^{-6}$ as compared with 0.059, 0.014, $3.55\ 10^{-3}$ in the original estimates for $y(1)$.

This idea of extrapolation can be used in two distinct ways. Firstly, as in the previous example at the end of the calculation to provided an improved solution at $x=1$. Secondly, the extrapolation can be made an integral part of the algorithm by including it at each step. If the solution of the differential equation is required over the interval $(0,1)$ at intervals of h we compute approximations for the solution at $x=h$ using a sequence of values $h, {}^{h}/_{2}, {}^{h}/_{4}, \ldots$ and then extrapolate as before. The extrapolated value is then used to produce the computed solution at $x=2h$, $3h, \ldots$. This idea forms the basis of many recent algorithms.

In general this type of extrapolation is dependent on knowing the form of the function g(h) given in equation (2.9.1). As a general rule of thumb if the method is of order p then the leading term of g will be proportional to h^p but this will depend upon the solution having a sufficiently smooth form. Rules of thumb are of course notoriously unreliable and there is no substitute for a formal analysis of the error whenever this is possible. If the function g(h) is smooth but p is unknown then it is still possible to extrapolate using the Aitken acceleration process. Let us assume that

$$y(x) - Y(x,h) = Ah^p, \tag{2.9.6}$$

where A and p are unknowns independent of h. Selecting step sizes h, $^h/_2$ and $^h/_4$ we shall also have that

$$y(x) - Y(x,{}^h/_2) = A({}^h/_2)^p, \tag{2.9.7}$$

and

$$y(x) - Y(x,{}^h/_4) = A({}^h/_4)^p. \tag{2.9.8}$$

This gives three equations in the unknowns y(x), A and p and in particular

$$y(x) = \frac{Y(x,h)\;Y(x,h/_4) - Y(x,h/_2)^2}{Y(x,h) - 2Y(x,h/_2) + Y(x,h/_4)} \tag{2.9.9}$$

Therefore, given Y(x,h), $Y(x,{}^h/_2)$ and $Y(x,{}^h/_4)$ we can extrapolate to improve the accuracy of the solution. If p is found then we can also investigate the order of convergence.

Example 2.9.2 From Example 2.9.1 an approximate solution of y'=y, subject to y(0)=1, at x=1 and using h=0.5 is Y(1,0.5)=2.77777777, likewise for h=0.25 and h=0.125 we have Y(1,0.25)=2.7326114 and Y(1,0.125)=2.7218319. Using equation (2.9.9) with h=0.5 gives the extrapolated value of $y(1)\simeq 2.7184527$ and $p\simeq 2.0669$ which confirms that this method is of second order.

Exercise 2.9.1 Use the results given in Table 2.2.1 to improve the accuracy of the computed solution.

Exercise 2.9.2 The following results were obtained when the Trapezium method was applied to the equation $y'=xy^2$, subject to $y(0)=1$: $Y(1,0.1)=2.02736$, $Y(1,0.05)=2.00661$ and $Y(1,0.025)=2.00164$. Confirm that this method is of order 2 and extrapolate the above results to give an improved estimate of the solution at $x=1$.

Exercise 2.9.3 Compute an approximation for the solution of $y'=\sqrt{x}\sin y$, subject to $y(0)=1$ using the Simple Runge-Kutta method given by equation (2.5.2) with $h=0.1,0.05,0.025$. Extrapolate the results obtained to improve the computed approximation for $y(1)$.

One final exercise demonstrates that the solution needs to be sufficiently smooth if we are to find an accurate solution, the existence of any singularity may limit the order of the method.

Exercise 2.9.4 Find a numerical solution of $y'=\sqrt{y}+\sqrt{x}$, subject to $y(0)=0$. Estimate the rate at which the Classical Runge-Kutta method converges using (2.9.6), (2.9.7) and (2.9.8). Use the results of your calculation to improve the computed solution.

2.10 Choice of Method

It is of course impossible to say which method is the "best". The answer to this question depends to a very large extent upon the problem being considered, the accuracy of solution which is required, the type of output needed, and the facilities in terms of software and hardware which are available. In this section a few ideas are given which might assist the choice of method employed to solve a given problem.

Predictor-Corrector methods are perhaps the most widely used numerical algorithms for solving first order differential equations.

Despite the problems inherent in multi-step methods they are relatively easy to implement and it is quite simple to estimate the local error at each step. The only major difficulty arises when it is shown to be prudent to reduce the step size, for then it will be necessary to interpolate previously computed values. This is usually accomplished by halving the step size and using a high order interpolation formula. (Obviously the order of the interpolation formula must exceed that of the difference method.) The major competitors to the predictor-corrector methods are the Runge-Kutta formulae which, since they are single step, do not suffer from the same problem if h is reduced. However, these methods may require numerous evaluations of f at each step. For example, the Merson variant requires 5 evaluations per step as against 2 for a predictor-corrector in PEC mode. There is also the problem of estimating the local discretization error when using Runge-Kutta formulae and even though this can be overcome to some extent using either the Merson or Fehlberg variants the estimates obtained are relatively crude. Nevertheless, the high order attainable by Runge-Kutta methods is very attractive.

If it is possible to estimate the local truncation error at each step then we can adjust the step size, either up or down, and this leads to the so called variable step methods. In general implicit methods are to be preferred since they will have higher order and a larger region of stability. This is offset by the requirement of the solution of a non-linear equation at each step. Reducing the step size h has two effects; firstly it reduces the local discretization error and secondly it will speed the convergence when solving the implicit recurrence relation to find Y_{n+1}. This is tempered by the fact that reducing h will increase the overall amount of work required to compute a solution and also increase the possibility of accumulating rounding errors. Instead of reducing h we may increase the order of the difference method which leads to the variable order variable step algorithms. Various criteria exist for determining when to change the step size h and/or order of the method, see for example Gear (1972) and Lambert (1973). There is much research in progress in developing automatic integrators. One of the first examples was proposed by Nordsieck (1962) which has a built in starting procedure and two tests, the first of which monitors the local error and the second attempts to examine stability. Nordsieck's method, despite its age, is a good general purpose method and is more economical than the general Runge-Kutta methods.

The practical control of error has two components, (i) an analysis which is carried out concurrently with the determination of the solution and (ii) a subsequent attempt to improve the computed result. For the former it is necessary to remember that it is the global error which we would like to control. This means that if we are trying to solve $y'=f(x,y)$, subject to $y(a)=\alpha$, then at some point $x=x_n$ we have to compute a solution of the local problem $y_n'=f(x,y_n)$, subject to $y_n(x_n)=Y_n$. At best we can try to make

$$|y_n(x_{n+1}) - y(x_{n+1})| \leqslant \tau,$$

where τ is a measure of the permissible local error, when in reality we wish to control the global error $|y(x_{n+1})-Y_{n+1}|$. Unfortunately, most procedures are incapable of this. As an example of a subsequent attempt to consider the growth of errors we have already looked at extrapolation or deferred approach to the limit and we have seen that provided the solution is sufficiently smooth that this can be very profitable. As an alternative we note that the solution of $y'=f(x,y)$, subject to $y(a)=\alpha$, is given by

$$y(x) = \alpha + \int_a^x f(x,y(x))dx. \qquad (2.10.1)$$

Since we have computed an approximation for the solution at a set of points $x_0=a,x_1,\ldots,x_n$ we can estimate the integral numerically and compare the results with $Y_n \simeq y(x_n)$. If there is a large discrepancy then it would appear that any step size or error tolerance must be reduced.

Example 2.10.1 The Classical Runge-Kutta method (2.5.5) was used to solve $y'=\sqrt{x}\ \sin y$, subject to $y(0)=1$, with $h=0.1$. The computed approximation for $y(1)=y(10\times h)\simeq Y_{10}$ was found to be 1.63193. Integrating, as for equation (2.10.1), using Simpson's rule gave $y(1)=1.6305$ which differs from Y_1 with relative error of 0.085%.

2.11 Higher Order and Systems of Equations of Initial Value Type

As it is always possible to reduce an N-th order differential equation to a system of N first order equations much of the previous five sections carries over directly.

Example 2.11.1 Consider the 3rd order equation

$$d^3y/dx^3 + p(x)d^2y/dx^2 + q(x)dy/dx + r(x)y = 0,$$

subject to the initial conditions $y(0)=\alpha_1$, $y'(0)=\alpha_2$, $y''(0)=\alpha_3$. We put $y=z_1$, $y'=z_2$ and $y''=z_3$ then

$$\begin{aligned} z_3' &= -p\ z_3 - q\ z_2 - r\ z_1, \\ z_2' &= z_3, \\ z_1' &= z_2. \end{aligned}$$

Now define $\mathbf{z} = (z_3,z_2,z_1)^T$ then we have the vector differential equation

$$\mathbf{z}' = \mathbf{A}\mathbf{z}, \qquad (2.11.1)$$

subject to $\mathbf{z}(0)=(y(0),y'(0),y''(0))^T$, where

$$\mathbf{A}(x) = \begin{bmatrix} -p(x) & -q(x) & -r(x) \\ 1 & 0 & 0 \\ 0 & 1 & 0 \end{bmatrix}.$$

The general 1st order system of equations may be written in the form

$$\mathbf{y}' = \mathbf{f}(x,\mathbf{y}), \qquad (2.11.2)$$

where $\mathbf{y}$ and f are vector functions of dimension N. The general solution of such an equation will involve exactly N arbitrary constants and to obtain a unique solution N additional conditions must be specified. If all N conditions are imposed at one point then the problem is said to be of **initial value type**, see Example 2.11.1. If the N conditions are imposed at more than one point the problem is said to be of **boundary value type**; such problems will be considered in the next chapter. The initial value problem $\mathbf{y}'=\mathbf{f}(x,\mathbf{y})$, subject to the initial condition $y(a)=\boldsymbol{\alpha}$, where y is a vector function of dimension N, has a unique solution provided f is continuous with respect to x in some interval containing x=a and Lipschitz with respect to $\mathbf{y}$ in $\Omega\subset R^N$; then for all x and for all vectors $\mathbf{y}$ and z in Ω there exists a scalar constant L>0 such that

$$\|\mathbf{f}(x,\mathbf{y}) - \mathbf{f}(x,\mathbf{z})\| \leq L \, \|\mathbf{y} - \mathbf{z}\|. \qquad (2.11.3)$$

For equation (2.11.1) such a condition is satisfied if $\|A\|$ is bounded but this is certainly the case if p,q and r are continuous functions on a closed bounded interval containing x=a for then they will also be bounded. This condition is sufficient to guarantee that equation (2.11.1) has a unique solution satisfying $\mathbf{y}(a)=\boldsymbol{\alpha}$.

Let us now consider whether such initial value problems are always well conditioned. As a simple example, consider the differential equation $y'' - 100y = 0$, subject to the initial conditions $y(0)=1$, $y'(0)=-1$, for which the solution is $y(x)={}^1/_{10}\exp(-10x)$. However, now consider the perturbed initial conditions $y(0)=1+\varepsilon$, $y'(0)=-1$ for which the solution is

$$y(x) = (1 + {}^{\varepsilon}/_{2})\exp(-10x) + {}^{\varepsilon}/_{2}\exp(10x),$$

which grows rapidly for any non-zero value of ε. Consequently the problem is very ill-conditioned; this instability is inherent within the problem and there is little or nothing that can be done about it. We see, therefore, that extreme caution must be exercised, as this type of perturbation is unavoidable.

We may develop numerical methods for equations such as (2.11.2) which are completely analogous to those produced in the previous sections. For example if $\mathbf{Y}_n=((Y_1)_n,(Y_2)_n,..,(Y_N)_n)^T$ is a computed approximation for $\mathbf{y}(x_n)=(y_1(x_n),y_2(x_n),..,y_N(x_n))^T$ then Euler's method for a system of differential equations $\mathbf{y}'=\mathbf{f}(x,\mathbf{y})$ is given by

$$\mathbf{Y}_{n+1} = \mathbf{Y}_n + h\mathbf{f}(x_n,\mathbf{Y}_n), \quad \mathbf{Y}_0 = \boldsymbol{\alpha}, \qquad (2.11.4)$$

In the same manner the Classical Runge-Kutta method (2.5.5) becomes

$$\mathbf{Y}_{n+1}= \mathbf{Y}_n+h[\mathbf{k}_1+2\mathbf{k}_2+2\mathbf{k}_3+\mathbf{k}_4]/6, \qquad (2.11.5)$$

where

$$\begin{aligned} \mathbf{k}_1 &= \mathbf{f}(x_n,\mathbf{Y}_n), \\ \mathbf{k}_2 &= \mathbf{f}(x_n+h/2,\mathbf{Y}_n+h/2\,\mathbf{k}_1), \\ \mathbf{k}_3 &= \mathbf{f}(x_n+h/2,\mathbf{Y}_n+h/2\,\mathbf{k}_2), \\ \mathbf{k}_4 &= \mathbf{f}(x_n+h,\mathbf{Y}_n+h\mathbf{k}_3). \end{aligned}$$

The predictor-corrector methods are extended in a similar way. We merely replace the scalar functions Y_n and f_n by their vector equivalents $\mathbf{Y}_n$ and $\mathbf{f}_n$.

Example 2.11.2 Find an approximate solution of the differential equation

$$d^2y/dx^2 = 2y^2/(1+x),$$

subject to $y(0)=1$, $y'(0)=-1$.

Solution. Putting y'=z the differential equation becomes

$$z' = 2y^2/(1+x), \qquad z(0)=y'(0)=-1,$$
$$y' = z, \qquad y(0)=1.$$

Applying Euler's method to these equations produces the recurrence relations

$$Z_{n+1} = Z_n + h\,2Y_n^2/(1+x_n),\ x_n=nh,\ Z_0=-1,$$
$$Y_{n+1} = Y_n + hZ_n, \quad Y_0=1.$$

If we take h=0.1 then $Y_{10}\approx y(10h)=y(1)$ and $Z_{10}\approx z(10h)=z(1)=y'(1)$. A little calculation gives $Y_{10}=0.5125784$ and $Z_{10}=-0.1735087$. Using h=0.05 produces $Y_{20}\approx y(20h)=y(1)$ and $Z_{20}\approx z(20h)=z(1)$ for which the computed values are $Y_{20}=0.5071041$ and $Z_{20}=-0.2117427$. The correct values of y(1) and y'(1) are 0.5 and -0.25 respectively. If the Classical Runge-Kutta method is applied to the problem then for h=0.1: $Y_{10}=0.5000206$, $Z_{10}=-0.2499759$, and for h=0.05: $Y_{20}=0.5000013$, $Z_{20}=-0.2499998$, showing a dramatic improvement at the expense of more effort per step.

Exercise 2.11.1 Confirm the above results by using the Classical Runge-Kutta method.

We have already seen that Euler's method produces an absolutely stable recurrence relation when applied to the test problem $y'=\lambda y$, $\mathrm{Re}(\lambda)>0$, only if $|1 + \lambda h|\leq 1$. As a corresponding test problem for systems of equations we use $\mathbf{y}'=\mathbf{A}\mathbf{y}$, where A is an N×N constant matrix. If A is similar to a diagonal matrix D, e.g. it has N distinct eigenvalues, then there is a a non-singular matrix S such that $\mathbf{S}\mathbf{A}\mathbf{S}^{-1}=\mathbf{D}$. Premultiplying the differential equation $\mathbf{y}'=\mathbf{A}\mathbf{y}$ by S then gives

$$\mathbf{S}\mathbf{y}' = \mathbf{S}\mathbf{A}\mathbf{S}^{-1}\,\mathbf{S}\mathbf{y}.$$

Setting $\mathbf{z}=\mathbf{S}\mathbf{y}$ reduces the above system to $\mathbf{z}'=\mathbf{D}\mathbf{z}$ which separates to give the equations $z_i'=\lambda_i z_i$, $1\leq n\leq N$, where λ_i is the diagonal entry of the ith row of D, or equivalently the ith eigenvalue of A. Therefore, the vector equivalent of Euler's method, (2.11.4), produces a stable vector recurrence relation provided $|1 + \lambda_i h| < 1$, $i=1,2,\dots,N$, where λ_i are the eigenvalues of **A**.

Example 2.11.3 Consider the system of equations

$$\mathbf{z}' = \begin{bmatrix} -50.5 & 49.5 \\ 49.5 & -50.5 \end{bmatrix} \mathbf{z}, \quad \mathbf{z}(0) = \begin{bmatrix} 0 \\ 2 \end{bmatrix} \qquad (2.11.6)$$

which has the solution $z_1 = e^{-x} - e^{-100x}$, $z_2 = e^{-x} + e^{-100x}$. Since both eigenvalues are real and negative the solution decays to zero, the term e^{-100x} vanishing very rapidly, and it might be expected that it is possible to ignore its contribution for relatively small values of x; however, this is not the case. The eigenvalues of the given matrix are -1 and -100 and so for Euler's method to produce an absolutely stable recurrence relation we require that $h < 2/100$ even for large values of x when the term exp(-100x) is effectively zero. Equations such as (2.11.6) are called **stiff** and were first discussed by C.W.Gear (1971). For an equation to be stiff it is necessary that all the eigenvalues have negative real part and that the so called stiffness ratio $S = \mathrm{Re}(-\lambda)_{max} / \mathrm{Re}(-\lambda)_{min} >> 0$. For this example S=100, but ratios of the order 10^6 are not uncommon in chemical kinetic problems.

Example 2.11.4 The differential equation $y' = -100(y-1)$ has a solution $y(x) = e^{-100x} + 1$ which rapidly decays to the value 1; however, the decaying transient solution cannot be ignored and a step size $h < 0.02$ is required for Euler's method to be stable.

For the general equation $\mathbf{y}' = \mathbf{f}(x, \mathbf{y})$ we consider the Jacobian matrix **A** given by $A_{ij} = \partial f_i / \partial y_j$ where f_i is the ith component of **f** and y_j the jth component of **y** and try to ensure that if λ_i is an eigenvalue of A then $h\lambda_i$, $i=1,2,\dots,N$, lies within the region of stability of the difference method being used. This can produce a very drastic restriction on the step size h, especially if the differential equation is in any way stiff. Let us suppose that f does not change too rapidly. Then as $h \to 0$ the values of $\lambda_i h$ tend to the origin within the wedges shown in Figure 2.11.1. Therefore, we can ensure stability provided the region of stability is given by $\pi-\alpha \leq \theta \leq -(\pi-\alpha)$. In which the method is said to be A(α)-stable. (See Widlund (1967).)

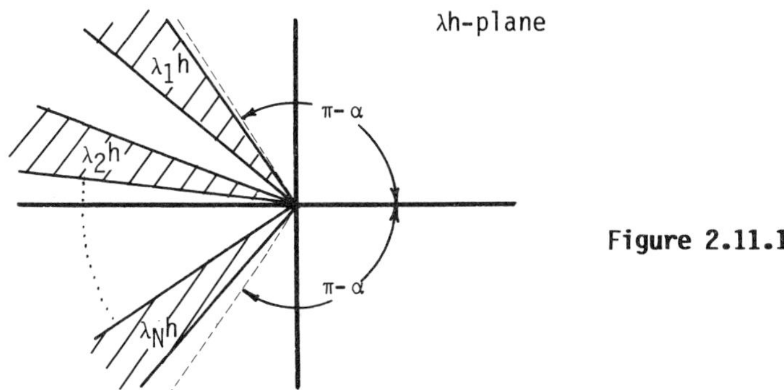

Figure 2.11.1

An alternative idea is given by Gear (1971) who defines the category of stiffly stable methods.

Definition 2.11.1 A method is said to be **stiffly stable** if when it is applied to the differential equation $y'=\lambda y$ it produces a recurrence relation which is (i) absolutely stable in the region $R_1=\{\operatorname{Re}(\lambda h) \leq D\}$ and (ii) accurate in the region $R_2=\{D<\operatorname{Re}(\lambda h)<\varepsilon, |\operatorname{Im}(\lambda h)|<\theta\}$.

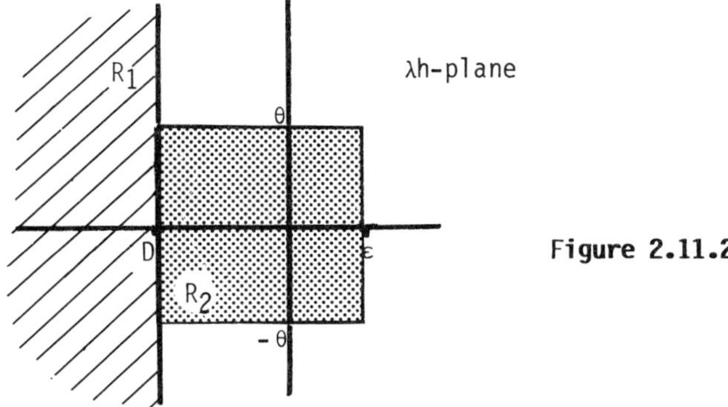

Figure 2.11.2

Figures 2.11.1 and 2.11.2 demonstrate the comparative regions of stability for an A(α)-stable and a stiffly stable method. The definition of stiffly stable methods is a little difficult to justify but we can argue as follows. Suppose that $\lambda h=u+iv$ then for $u<D<0$ the solution is reduced by at most e^u per step and as we are not interested in the accuracy of small components we ignore these components in R_1. Near the origin we are interested in accuracy and so we shall require both absolute and relative stability. If $u>\varepsilon>0$ then the difference solution increases by at most e^ε per step and clearly the size of ε must be limited. Therefore, if as f varies its Jacobian has eigenvalues which drift into the positive half plane for some arguments then we shall still have stability provided their real parts are less than ε. Gear (1971) gives further details regarding the limits on both ε and θ.

Gear has shown that in order for a method to be stiffly stable the degree of σ must be at least as large as that of ρ in which case the method will be implicit. Such methods will require the solution of a non-linear equation at each step which can be solved using the simple iteration

$$\{Y_{n+1}\}^{p+1} = h\beta_0 \mathbf{f}(x_n,\{Y_{n+1}\}^p)+c_n$$

which converges if $|h\beta_0 L|<1$, where L is the Lipschitz constant associated with **f**. However, if the equations are stiff L will be very large and so h needs to be small to ensure convergence. As an alternative it is possible to use a Newton type iteration to find $\mathbf{Y}_{n+1}$ but this will in general require the calculation of the Jacobian at each step; a very time consuming task. Instead we can use an iteration in which the Jacobian is only computed when either the step size is changed or else the solution is undergoing rapid variations. (See for example Gear (1971) or Hall and Watt (1976).)

It is possible to find stiffly stable methods for $m=1,2,3,4,5,6$ with $\sigma(r)=\beta_0 r^m$; but $m=7$ gives an unstable formula. The corresponding values of $\alpha_0,\ldots,\alpha_m$ are given in Table 2.11.1. Jain (1979) gives details of alternative stiffly stable methods which can be obtained using $\sigma(r)=\beta_0 r^{m-j}(r-\gamma)^j$, $j=0,1,2,\ldots,m$ where $-1\leqslant\gamma<1$; this gives methods of order $\leqslant j+6$ if $j<5$ and of order $\leqslant 11$ if $j=6$. There are no stiffly stable methods of order in excess of 12.

Table 2.11.1 Stiffly Stable Methods for the choice $\sigma(r)=\beta_0 r^m$.

m	β_0	α_0	α_1	α_2	α_3	α_4	α_5	α_6	order
1	1	1	-1						1
2	2	3	-4	1					2
3	6	11	-18	9	-2				3
4	12	25	-48	36	-16	3			4
5	60	137	-300	300	-200	75	-12		5
6	60	147	-360	450	-400	225	-72	10	6

The polynomial $\rho(r)$ is given by $\alpha_0 r^m + \alpha_1 r^{m-1} + \ldots + \alpha_m$.

Example 2.11.5 The regions of absolute stability of the methods given in Table 2.11.1 are given by Figure 2.11.3. (The region of stability is to the left of the given locus.)

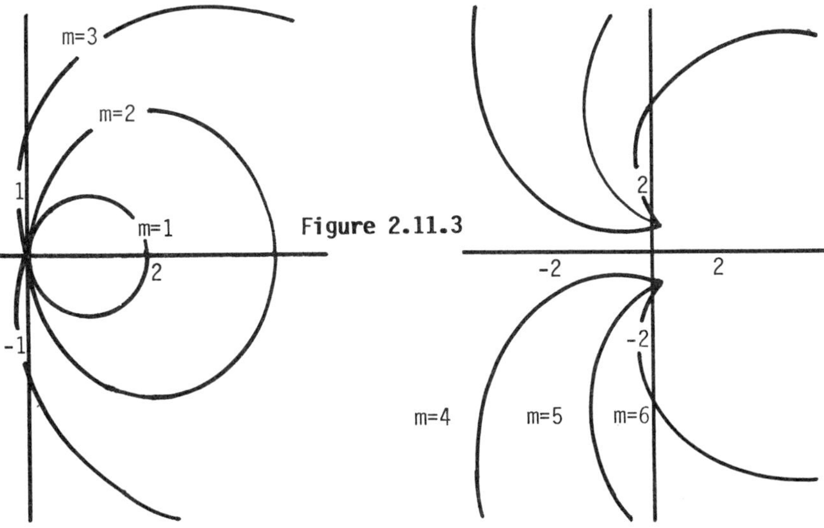

Figure 2.11.3

The Backward Euler method is A-stable and is therefore $A(\pi/2)$-stable and stiffly stable with D=0. The stiffly stable method of order 6, given in Table 2.11.1, is not A-stable but it is $A(\alpha)$-stable with $\alpha \simeq 26°$ and stiffly stable with $D \simeq -6$, $\theta \simeq 1/2$ and $\varepsilon = 0$.

Exercise 2.11.2 Show that the regions of absolute stability for the methods given in table 2.11.1 are as given in Figure 2.11.3.

Exercise 2.11.3 Show that the order of the stiffly stable method with step number m given in Table 2.11.1 is m.

Exercise 2.11.4 Find an approximate solution of the equations $\mathbf{y}'=\mathbf{A}\mathbf{y}+\mathbf{b}$ where

$$A = \begin{bmatrix} -1999+3i & 999-2i \\ -1998+4i & 998-3i \end{bmatrix}, \qquad \mathbf{b} = \begin{bmatrix} 1 \\ 1 \end{bmatrix},$$

subject to $y(0)=(2,2)^T$ using a suitably stable difference method.

Exercise 2.11.5 (Gear 1971) Find an approximate solution of the equations $\mathbf{y}'=\mathbf{f}(x,\mathbf{y})$, where $f_i(y_1,y_2,y_3,y_4)=-\beta_i y_i+(y_i)^2$, with $\beta_1=1000$, $\beta_2=800$, $\beta_3=-10$ and $\beta_4=0.0001$, subject to $y_i(0)=-1$. For which values of x would these equations be considered stiff? (The general solution of the given equation is $y_i(x)=\beta_i/[1+c_i\exp(\beta_i x)]$.)

2.12 Special Methods for Initial Value Problems of Second Order

Though it is possible to reduce the second order equation

$$y'' = f(x,y,y'), \qquad (2.12.1)$$

subject to the initial conditions $y(a)=\alpha_1$ and $y'(a)=\alpha_2$, to the standard form (2.11.2) it is sometimes more convenient to consider this equation directly, especially when f is independent of y'. In order to approximate the equation $y''=f(x,y)$ we need some means of representing $y''(x)$ in terms of the solution at x and the neighbouring points x+h and x-h. We can do this as follows.

If the solution has sufficiently many derivatives then

$$y(x+h)=y(x)+hy'(x)+0.5h^2y''(x)+O(h^3)$$

and

$$y(x-h)=y(x)-hy'(x)+0.5h^2y''(x)+O(h^3).$$

Adding these two expansions gives

$$y(x+h)+y(x-h)=2y(x)+h^2y''(x)+O(h^4),$$

therefore, we can approximate y'' by $(y(x+h)-2y(x)+y(x-h))/h^2$ to give

$$\frac{y(x+h) - 2y(x) + y(x-h)}{h^2} = f(x,y(x)) + E(h), \qquad (2.12.2)$$

where $E(h)$ is the local discretization error. If h is sufficiently small we neglect $E(h)$ and obtain the difference equation

$$\frac{Y_{n+1} - 2Y_n + Y_{n-1}}{h^2} = f(x_n,Y_n), \qquad (2.12.3)$$

From equation (2.12.2) the local discretization error, $E(h)$, is given by $h^2/12\ d^4y/dx^4(\xi_n)$, where ξ_n lies between $x-h$ and $x+h$, and so equation (2.12.3) is consistent of order 2 with the differential equation (2.12.1). A more accurate scheme can be obtained by considering the general method

$$\frac{Y_{n+1} - 2Y_n + Y_{n-1}}{h^2} = \beta_0 f_{n+1} + \beta_1 f_n + \beta_2 f_{n-1}, \qquad (2.12.4)$$

where $f_p=f(x_p,Y_p)$.

Exercise 2.12.1 Show that if $\beta_0=\beta_2=1/12$, $\beta_1= 10/12$ the difference method (2.12.4) is consistent of order 4. This scheme is called Numerov's method or the Royal Road formula.

The difference method (2.12.2), and in general (2.12.4), will produce a second order recurrence relation and will therefore require two starting values Y_0 and Y_1 say. However, the given initial conditions are in terms of y(a) and y'(a) and so it is necessary to estimate Y_1 as accurately as possible before the recurrence relation can be begun. We set $Y_0=y(a)$ and then obtain Y_1 as

$$Y_1 \simeq y(x_1) = y(a+h) = y(a) + hy'(a) + {}^{h^2}/_2 y''(a) + O(h^3),$$

$$= y(a) + hy'(a) + {}^{h^2}/_2 f(a,y(a),y'(a)) + O(h^3). \quad (2.12.5)$$

If Numerov's method, given in Exercise 2.12.1, is used then it is necessary to use at least the first 5 terms in the expansion of y(x+h).

In general the difference method (2.12.4) will be implicit and so it will be necessary to solve what will usually be a non-linear equation at each step. As the methods which are normally applied are very similar to those applied to first order equations no further details will be given here. The errors caused by only applying a finite number of iterations when solving such equations and also due to rounding errors can be examined as before.

Exercise 2.12.2 Show that for Numerov's method, given by Exercise 2.12.1, the iteration

$$\{Y_{n+1}\}^{r+1} = {}^{h^2}/_{12} f(x_{n+1}, \{Y_{n+1}\}^r) + C_n, \quad \{Y_{n+1}\}^0 = Y_n,$$

where $C_n=2Y_n-Y_{n-1}+h^2[10f(x_n,Y_n)+f(x_{n-1},Y_{n-1})]/_{12}$, converges provided $|h^2 {}^{\partial f}/_{\partial y}| < 12$.

As the recurrence relation (2.12.4) is of the same order as the differential equation there will be no parasitic solutions although it is still possible for numerical instabilities to arise.

Example 2.12.1 Show that the difference method

$$Y_{n+1}-2Y_n+Y_{n-1} = h^2f(x_n,Y_n)$$

when applied to the test problem $y''=-\lambda^2y$, $\lambda>0$, produces a second order recurrence relation which is absolutely stable provided $\lambda h<2$.

Solution. The general solution of $y''=-\lambda^2y$ is given by the bounded function $y(x)=A\cos\lambda x+B\sin\lambda x$, where A and B are arbitrary constants. In contrast, the difference solution is computed using the recurrence relation

$$Y_{n+1}-2Y_n+Y_{n-1} = -\lambda^2h^2Y_n$$

i.e.

$$Y_{n+1}-(2-\lambda^2h^2)Y_n+Y_{n-1}=0.$$

The solution of this equation is given by $Y_n=Cr^n+Dr^n$ where r_1 and r_2 are the roots of $r^2-(2-\lambda^2h^2)r+1=0$ and C and D are arbitrary. If the roots are real then, since the product r_1r_2 is equal to 1, either $|r_1|>1$ or $|r_2|>1$ and the solution is unbounded. Therefore, we require that r_1 and r_2 are a complex conjugate pair; and so, for stability it is necessary that $(2-\lambda^2h^2)^2\leq4$, i.e. $\lambda h\leq2$.

Exercise 2.12.3 Show that Numerov's method, given in Exercise 2.12.1, is stable for the test problem $y''=-\lambda^2y$ provided that $\lambda^2h^2\leq6$.

Exercise 2.12.4 Find an approximate solution of $y''=-10\sin y$ subject to $y(0)=0$, $y'(0)=0.1$. This is the equation of motion of a simple pendulum which passes through its datum with velocity 0.1. If the amplitude of the oscillation remains small then the above equation may be approximated by $y''=-10y$ for which $y(x)=-\sqrt{10}\cos(x)$. Is the assumption of small oscillations valid?

Provided the procedure for determining Y_1 given by equation (2.12.5), or an equivalent, is performed the difference solution converges to the solution of the differential equation (2.12.1).

Theorem 2.12.1 Let the difference method

$$Y_{n+1} = 2Y_n - Y_{n-1} + h^2[\beta_0 f_{n+1} + \beta_1 f_n + \beta_2 f_{n-1}] \qquad (2.12.6)$$

be consistent of order k with the differential equation $y''=f(x,y)$, where $x_n=a+nh$ and $f_n=f(x_n,Y_n)$. Furthermore, let $Y_0=y(0)$ and

$$|Y_1 - y(x_1)| \leqslant M h^q, \; q \geqslant k, \text{ M a positive constant,}$$

then the computed difference solution is convergent of order k.

As with all numerical methods if it is possible to obtain an expression for the computational error then we may extrapolate to improve the calculated solution. However, this technique must be treated with care as was shown by Fox (1962). Consider the following example.

Example 2.12.2 Find a numerical solution of the differential equation $y''=12x^2$, subject to $y(0)=y'(0)=0$ and compare it with the exact solution $y(x)=x^4$.

Solution. For this example the difference method (2.12.2) reduces to the linear second order recurrence relation

$$Y_{n+1}-2Y_n+Y_{n-1}=12h^2x_n^2=12h^4n^2,$$

since $x_n=nh$. This equation can be solved using the techniques of section 1.2. The general solution is $Y_n=h^4(n^4-n^2)+an+b$ and since $Y_0=y(0)=0$ we have that b=0. We can use equation (2.12.5) to give

$$Y_1 \simeq y(h) = y(0) + hy'(0) + {}^{h^2}/_2\, y''(0) + \ldots\ldots = 0 + O(h^4),$$

therefore, a=0 and the solution is $Y_n=x_n^4-h^2x_n^2$. The error between Y_n and $y(x_n)=x_n^4$ is $O(h^2)$ and we can apply Romberg extrapolation to produce the exact answer in one iteration. As an alternative to approximating Y_1 using equation (2.12.5) if we use the exact value for $Y_1=y(h)=h^4$ the calculated solution is slightly better. (See Table 2.12.1.) However, if we extrapolate on the assumption that since the discretization error involves only even powers we need only eliminate h^2, h^4,.. the final results are worse than using the approximate value $Y_1=0$.

From Table 2.12.1 using h=0.5 and h=0.25 produces approximate values for y(1) of 0.875 and 0.953125. These give the extrapolated value

$$0.953125 + {}^{(0.953125-0.875)}/_{15} = 0.9791666$$

which can be compared with the correct value of y(1)=1. The explanation of this is as follows. From the general solution of the difference equation we get b=0 but $Y_1=h^4$ gives $a=h^4$ and then $Y_n=x_n^4-h^2x_n + h^3x_n$. The presence of a term $O(h^3)$ means that any attempt to extrapolate using only even powers cannot succeed.

Table 2.12.1 Results of applying Romberg Extrapolation to $y''=12x^2$.

	(i) $Y_1=0$, (ii) $Y_1=h^4$.				
(i)	x=0	x=0.25	x=0.50	x=0.75	x=1.00
h=0.50	0		0		0.75
h=0.25	0	0	0.046875	0.28125	0.9375
(ii)					
h=0.50	0		0.0625		0.875
h=0.25	0	0.003906	0.054688	0.292969	0.953125
y(x)	0	0.003906	0.062500	0.316406	1.000000

Exercise 2.12.5 Compute a numerical solution of $y''=-y$ subject to $y(0)=0$, $y'(0)=1$ using (i) the difference method (2.12.2), and (ii) the Royal Road formula with h=0.1,0.05,0.025. Confirm that the method (i) is $O(h^2)$ and (ii) is $O(h^4)$ for this example. Extrapolate to obtain an improved solution in each case.

Exercise 2.12.6 Compute a numerical solution of $y''=2y^3$ subject to $y(0)=1$, $y'(0)=-1$ using (i) the difference method (2.12.2), (ii) the Royal Road formula. Compare the computational effort of these methods with the technique of reducing the differential equation to a first order system and applying (i) the Simple Runge-Kutta and (ii) the Classical Runge-Kutta method.

CHAPTER 3

Ordinary Differential Equations—Boundary Value Problems

3.1 Introduction

As the general solution of either an N-th order differential equation or system of N first order equations involves N arbitrary constants, N additional conditions are needed to obtain a particular solution from the general solution. If such conditions are specified at more than one point then the problem is said to be of **boundary value type.**

Example 3.1.1 Find the solution of the second order differential equation $y''+\lambda^2y=1$, subject to $y(0)=0$, $y(1)=1$.

Solution. The general solution of the homogeneous equation $y''+\lambda^2y=0$ is $y(x)= A\sin\lambda x + B\cos\lambda x$, where A and B are arbitrary constants. The function $y(x)={}^1/_{\lambda^2}$ is a particular solution of the inhomogeneous equation, therefore, since the given differential equation is linear its general solution is

$$y(x)= A\sin\lambda x + B\cos\lambda x + {}^1/_{\lambda^2}. \qquad (3.1.1)$$

In order to satisfy the boundary condition $y(0)=0$ we require that $B=-1/\lambda^2$ and then the value of A which ensures that $y(1)=1$ is given by

$$A = (\lambda^2+\cos\lambda-1)/\lambda^2\sin\lambda. \qquad (3.1.2)$$

With an initial value problem it is relatively easy to determine whether a particular differential equation has a unique solution passing through a given point; for a boundary value problem the situation is not so simple. Though the differential equation alone may have a solution it is quite possible that certain boundary conditions yield no solutions, a unique solution or even an infinite number of solutions.

Example 3.1.2 The differential equation

$$y'' + \pi^2 y = 0, \text{ subject to } y(0)=0,\ y(1)=\varepsilon,\ \varepsilon\neq 0, \qquad (3.1.3)$$

has no solutions, but

$$y'' + \pi^2 y = 0, \text{ subject to } y(0)=0,\ y(1)=0, \qquad (3.1.4)$$

has an infinite number of solutions. On the other hand, the same equation subject to $y(0)=0$ and $y(1/2)=1$ has a unique solution given by $y(x)=\sin(\pi x)$. Conditions which are sufficient to guarantee a unique solution have been obtained only for certain limited classes of problems.

Theorem 3.1.1 The differential equation

$$y'' = f(x,y), \text{ subject to } y(a)=\alpha,\ y(b)=\beta, \qquad (3.1.5)$$

has a unique solution on the interval $a\leq x\leq b$ if $\partial f/\partial y>0$.

For more general results see Coddington and Levinson (1955), Collatz (1960), Keller (1968) or Daniel and Moore (1970). The above result is of only limited practical use if f is non-linear, for it may be necessary to find y in order to check whether the condition is satisfied.

Example 3.1.3 Consider the differential equation

$$y'' = y^2, \text{ subject to } y(0)=1,\ y(1)={}^1/_2. \qquad (3.1.6)$$

For this equation $\partial f/\partial y = 2y$ and so we can guarantee a unique solution if $y>0$ in $(0,1)$ but we have no way of checking this before the solution is computed.

Even if a solution of the problem (3.1.5) exists there is no guarantee that it will be well conditioned. For example, the solution given by equation (3.1.3) is well conditioned provided λ is not close to a multiple of π for if it is then $\sin(\lambda)\approx 0$ and so $A\to\infty$. In this chapter we shall examine some of the techniques available for computing solutions for differential equations of boundary value type. For simplicity we shall concentrate on second order equations, but much extends without a great deal of difficulty to higher order differential equations.

3.2 Simple Shooting Methods

We begin by considering the linear two point boundary value problem

$$y'' = p(x)y' + q(x)y + r(x), \quad y(0)=\alpha,\ y(1)=\beta,$$

and show that by exploiting the linearity of the differential equation it is possible to construct a solution which satisfies the given boundary conditions by solving two closely associated initial value problems.

Let v(x) satisfy the inhomogeneous differential equation

(i) $v''=q(x)v'+p(x)v+r(x)$, subject to $v(a)=\alpha$ and $v'(a)=0$,

and u(x) satisfy the homogeneous equation

(ii) $u''=q(x)u'+p(x)u$, subject to $u(a)=0$ and $u'(a)=1$,

then the function $y(x)=v(x)+\lambda u(x)$ satisfies the inhomogeneous differential equation and $y(a)=\alpha$. We now select the parameter λ so that $y(b)=v(b)+\lambda u(b)=\beta$ or $\lambda={}^{[\beta-v(b)]}/_{u(b)}$. This approach can be compared with the solution of the second order recurrence relation given in Example 1.2.3.

We illustrate the simple shooting method with an example.

Example 3.2.1 Use the simple shooting method to find a numerical solution of the equation

$$y'' - y = x - 2, \tag{3.2.1}$$

subject to $y(0)=2$, $y(1)=1$.

Solution. Consider differential equations

(i) $v'' - v = x-2$, subject to $v(0)=2$, $v'(0)=0$, (3.2.2)

and

(ii) $u'' - u = 0$, subject to $u(0)=0$, $u'(0)=1$, (3.2.3)

and then set $y(x)=v(x)+\lambda u(x)$. The function y(x) satisfies the equation (3.2.1), furthermore, at x=0: $y(0)=v(0)+\lambda u(0)=2$ and at x=1: $y(1)=v(1)+\lambda u(1)$. If v(1) and u(1) were known we could then select a value of λ such that $y(1)=1$, i.e. $\lambda={}^{[1-v(1)]}/_{u(1)}$, and then $y(x)=v(x)+\lambda u(x)$ is the required solution. We can solve both equation (3.2.2) and equation (3.2.3) using any of the methods developed in the previous chapter since they are of initial value type; to illustrate this method we apply the Classical Runge-Kutta method with step size h=0.1 to the vector equivalents of (3.2.2) and (3.2.3). (See Chapter 2, section 11.) This produces approximations for $v(x_n)$ and $u(x_n)$, $x_n=nh$, which will be denoted by V_n and U_n respectively. In particular the computed approximations for v(1) and u(1) are $V_{10}=2.1751999$ and $U_{10}=1.1751999$ producing a value $\lambda=-1$. We then obtain $Y_n=V_n+\lambda U_n$, n=0,1,..,10, giving the results in Table 3.2.1.

Table 3.2.1 Results of applying the Simple Shooting Method to $y''-y=x-2$, subject to $y(0)=2, y(1)=1$.

	n=2	n=4	n=6	n=8	n=10
x_n	0.2	0.4	0.6	0.8	1.0
$V_n \approx v(x_n)$	2.00133583	2.01075198	2.03665301	2.0881051	2.1751999
$U_n \approx u(x_n)$	0.20133583	0.41075194	0.63665301	0.8881051	1.1751999
$Y_n \approx y(x_n)$	1.80000000	1.60000000	1.40000000	1.2000000	1.0000000

Exercise 3.2.1 Use the simple shooting method to find an approximate solution of the linear two point boundary value problem given by $x^2y''-2y=2$, subject to $y(1)=1$ and $y(2)=0$.

Exercise 3.2.2 The results given in Table 3.2.1 are exact. Why? (Hint: Look at the local discretization error of the Classical Runge-Kutta method.)

Though the simple shooting method looks very attractive it can suffer from severe induced instability.

Example 3.2.2 Use the simple shooting method to find a solution of $y''=30^2(y-1+2x)$, subject to $y(0)=1$ and $y(1)=-1$.

Solution. We solve $v''=30^2(v-1+2x)$, subject to $v(0)=1, v'(0)=0$, and $u''-30^2u=0$ subject to $u(0)=0, u'(0)=1$, using the Classical Runge-Kutta method with h=0.01. Since 100h=1 the computed approximation for v(1) and u(1) will be V_{100} and U_{100} respectively. A little calculation gives $V_{100}=3.55654\times10^{11}$ and $U_{100}=1.77827\times10^{11}$; therefore, $\lambda=(-1-1.77827\times10^{11})/3.55654\times10^{11} = -0.5$. Notice that the boundary condition at x=1 has been completely lost; in fact, working to 8 significant figures any value for y(1) between -100 and 100 will be ignored. Clearly, this induced error needs to be examined in detail. The general solution of the given differential equation is $y(x)=Ae^{30x}+Be^{-30x}+1-2x$ and to satisfy the boundary conditions $y(0)=1, y(1)=-1$ it is necessary that both A and B are zero. This choice for A and B gives $y'(0)=-2$; any other choice will introduce the rapidly increasing component e^{30x} into the solution and this term soon swamps the required solution. Notice that if we start at x=1 and integrate backwards, i.e. replace x by 1-x, then the component e^{-30x} becomes dominant and this term will swamp the required solution.

With second order equations at least one condition must be provided at either end of the interval and so there is no a priori evidence as to the preferred direction in which to integrate the differential equation. The situation with higher order equations is not quite so obvious.

Example 3.2.3 Compute an approximate solution of the differential equation

$$x^3 \, d^3y/dx^3 + x \, dy/dx - y = -3+\log x,$$

subject to $y(1)=1$, $y'(2)=1/2$, $y''(2)=1/4$.

Solution. Firstly, we reduce this 3rd order equation to a system of 3 first order equations by putting $y=z_1$, $y'=z_2$ and $y''=z_3$. This gives

$$\mathbf{z}'=\mathbf{A}\mathbf{z}+\mathbf{b},$$

where

$$\mathbf{A} = \begin{bmatrix} 0 & 1 & 0 \\ 0 & 0 & 1 \\ x^{-3} & -x^{-2} & 0 \end{bmatrix}, \quad \mathbf{b} = \begin{bmatrix} 0 \\ 0 \\ [-3+\log x]/x^3 \end{bmatrix},$$

subject to the boundary conditions that at x=1: $z_1(1)=1$ and at x=2: $z_2(2)=1/2$ and $z_3(2)=1/4$. If we integrate from x=1 towards x=2 then we will need to guess both $z_2(1)$ and $z_3(1)$ before we can begin, but starting at x=2 and integrating back towards x=1 we need only estimate $z_1(2)$ and so the process is much simpler. However, we shall use the former method to demonstrate how such a solution can be found when more than one parameter is needed.

As initial values for z_2 and z_3 are unknown we take $z_2(1)=\alpha$ and $z_3(1)=\beta$, i.e. $z = (1,\alpha,\beta)^T$ and use this to produce a solution for values x between x=1 and x=2. We then exploit the linearity of the differential equation and note that if we solve

(i) $\mathbf{u}'=\mathbf{Au}+\mathbf{b}$ subject to $u(1)=(1,0,0)^T$,
(ii) $\mathbf{v}'=\mathbf{Av}$ subject to $v(1)=(0,1,0)^T$,
(iii) $\mathbf{w}'=\mathbf{Aw}$ subject to $w(1)=(0,0,1)^T$,

then $z(x)=\mathbf{u}(x)+\alpha\mathbf{v}(x)+\beta\mathbf{w}(x)$ satisfies the inhomogeneous equation $\mathbf{z}'=\mathbf{Az}+\mathbf{b}$, subject to $z(1)=\mathbf{u}(1)+\alpha v(1)+\beta\mathbf{w}(1)=(1,\alpha,\beta)^T$. At x=2

$$\begin{bmatrix} z_1(2) \\ z_2(2) \\ z_3(2) \end{bmatrix} = \begin{bmatrix} u_1(2)+\alpha v_1(2)+\beta w_1(2) \\ u_2(2)+\alpha v_2(2)+\beta w_2(2) \\ u_3(2)+\alpha v_3(2)+\beta w_3(2) \end{bmatrix}$$

and we require values for α and β such that $y'(2)=z_2(2)=1/2$ and $y''(2)=z_3(2)=1/4$. Therefore, to satisfy the boundary conditions at x=2 we select α and β according to

$$\begin{bmatrix} v_2(2) & w_2(2) \\ v_3(2) & w_3(2) \end{bmatrix} \begin{bmatrix} \alpha \\ \beta \end{bmatrix} = \begin{bmatrix} z_2(2)-u_1(2) \\ z_3(2)-u_2(2) \end{bmatrix} \qquad (3.2.4)$$

We solve the problems (i), (ii) and (iii) using the Classical Runge-Kutta method with h=0.01 and compute approximate values for $\mathbf{u}$, $\mathbf{v}$ and $\mathbf{w}$. Let U_n, V_n and W_n be the computed approximations for $\mathbf{u}(x_n)$, $v(x_n)$ and $w(x_n)$, $x_n=1+nh$, then at x=2=1+100h we obtain

$$U_{100} = \begin{bmatrix} 0.82639984 \\ -0.43337369 \\ -0.59657360 \end{bmatrix}, \quad V_{100} = \begin{bmatrix} 0.905841346 \\ 0.759773491 \\ -0.346575593 \end{bmatrix}, \quad W_{100} = \begin{bmatrix} 0.480453014 \\ 0.933373687 \\ 0.846573588 \end{bmatrix}$$

and equation (3.2.4) becomes

$$\begin{bmatrix} 0.759773491 & 0.933373687 \\ -0.346573593 & 0.846573588 \end{bmatrix} \begin{bmatrix} \alpha \\ \beta \end{bmatrix} = \begin{bmatrix} 0.933373687 \\ 0.846573588 \end{bmatrix}$$

from which $\alpha=1$ and $\beta=0$. The solution is then given by $Y_n=U_n+\alpha V_n+\beta W_n$, $n=0,1,..,100$.

Exercise 3.2.3 Confirm the results of Example 3.2.3.

Exercise 3.2.4 Solve the problem given in Example 3.2.3 by integrating backwards from $x=2$ towards $x=1$.

The situation with non-linear boundary value problems is a little more complicated. However, we can still use the same idea of estimating any unknown parameters and then attempt to systematically improve such values. We illustrate the approach with an example.

Example 3.2.4 Consider the following non-linear two point boundary value problem

$$y''= {}^{2y^2}/_{(1+x)}, \quad y(0)=1, \ y(1)=1. \qquad (3.2.5)$$

Solution. As before we set $z=y'$ and write the differential equation as

$$\begin{aligned} z' &= {}^{2y^2}/(1+x), \\ y' &= z, \qquad y(0)=1. \end{aligned} \qquad (3.2.6)$$

Since these equations are non-linear we cannot superpose solutions as in Example 3.2.1. Furthermore, in order to integrate (3.2.6) as an initial value problem we require a value for $z(0)$ but no such value is given. However, if we take a guess for z(0) and use it to compute a numerical solution we can then compare the calculated value for y at x=1 with the given boundary condition y(1)=1 and adjust the estimated value, z(0), to give a better approximation for the solution. Since the derivative at x=0 gives the trajectory of the computed solution this technique is called a **shooting method.** For example, if we take a series of values for z(0) and apply the Classical Runge-Kutta method with h=0.1 this procedure produces the results shown in Figure 3.2.1.

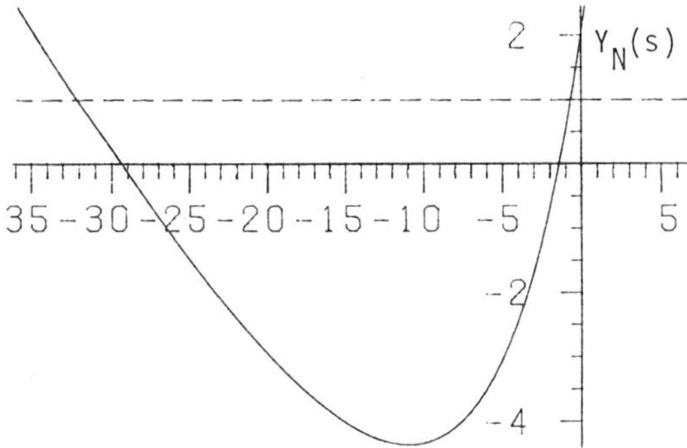

Figure 3.2.1

The computed value for y(1) using an initial estimate for z(0)=s.

We see from Figure 3.2.1 there are two possible values for z(0) which give y(1)=1. One value of z(0) is in the interval (-32.5,-32) and the other in the interval (-1.0,-0.5). We consider the second solution a little further. Taking various values for z(0) between -1 and -0.5 we compute a corresponding difference solution using a step size h=0.01. The results are given in Table 3.2.2.

Table 3.2.2 Computed solution of $y''=2y^2/(1+x)$, y(0)=1, y'(0)=z(0)=s.

z(0)	Computed solution at x=1
-1.00	0.5000202
-0.90	0.6364689
-0.80	0.7759338
-0.70	0.9184760
-0.60	1.0641581
-0.50	1.2130445

Since we seek a solution which satisfies $y(1)=1$ the required value of $z(0)$ lies between -0.70 and -0.60. By successively refining the interval a suitable solution can be found. It is also possible to improve the solution by linear interpolation. In Figure 3.2.2 the computed values for the solution at $x=1$ using $z(0)=-0.7$ and $z(0)=-0.6$ have been plotted. Next we determine where the straight line joining these points cuts the axis, i.e. $-0.7+\alpha$. This gives $\alpha=0.0559602$ and hence we try $z(0)=-0.7+\alpha=-0.6440397$. This value is then used to compute a new approximation; in particular the computed approximation $y(1)$ is 0.999599. We can then repeat this process. Alternatively, by fitting a suitable curve to the data given in Table 3.2.2 it is possible to obtain a more accurate estimate for $z(0)$. It is also possible to use inverse interpolation. (See Fox (1962) or Noble (1966).) However, it is more efficient to use a more automatic approach.

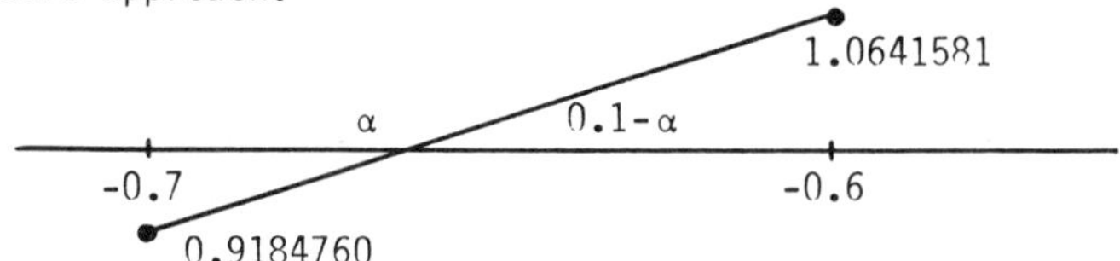

Figure 3.2.2 Linear Interpolation Applied to Example 3.2.4.

Let $Y_n(s), Z_n(s)$, $n=0,1,\ldots,N$ be a computed solution for the equation (3.2.6) which is obtained by setting $Y_0=y(0)$ and $Z_0=z(0)=s$. As $Nh=1$, $Y_N(s)$ is the computed solution for $y(1)$ and we require a value of s such that $Y_N(s)=y(1)$, i.e. we need to find the zeros of the function $F(s)=Y_N(s)-y(1)=0$. This can be accomplished by using Newton's method in the form

$$s_{p+1} = s_p - F(s_p)/F'(s_p), \qquad (3.2.7)$$

where $F'(s)=\frac{d}{ds}(F(s))$. Notice that $F'(s)=\frac{d}{ds}(Y_N(s)-1)=\frac{\partial Y_N}{\partial s}$ and so $\frac{\partial Y_N}{\partial s}$ satisfies

$$\frac{\partial}{\partial s}(y'')=\left(\frac{\partial y}{\partial s}\right)''=\frac{\partial}{\partial s}(f(x,y,y')) = \frac{\partial f}{\partial y}\frac{\partial y}{\partial s} + \frac{\partial f}{\partial y'}\frac{\partial y'}{\partial s}.$$

If we put $v={}^{\partial y}/_{\partial s}$ then the function v satisfies the differential equation $v''=f_y v+f_{y'}v'$ subject to the boundary conditions $v(0)={}^{\partial}/_{\partial s}(y(0))=0$ and $v'(0)={}^{\partial}/_{\partial s}(y'(0))={}^{\partial}/_{\partial s}(s)=1$. For Example 3.2.4 we have $f=2(1+x)^{-1}y^2$, $f_y=4(1+x)^{-1}y$ and $f_{y'}=0$ and so v satisfies $v''=2(1+x)^{-1}yv$, subject to $v(0)=0$ and $v'(0)=1$. Therefore, in order to integrate the differential equation (3.2.6) we estimate z(0) to be s_0 and then proceed as follows. We solve the differential equation $y''=f(x,y)$, subject to $y(0)=1$ and $y'(0)=s_0$ to find $Y_1,Y_2,\ldots,Y_N$ as before. Next we solve $v''=4(1+x)^{-1}yv$, subject to $v(0)=1$ and $v'(0)=1$ to get $V_1,V_2,\ldots,V_N$ and update s_p by

$$s_{p+1} = s_p - [Y_N(s_p) - 1]/V_N.$$

Notice that the differential equation for v involves the current approximation for the solution and so we need to replace $y(x_n)$ by Y_n as it is calculated. Beginning with $s_0=-0.65$ this produces the following sequence of estimates for z(0).

Iteration p	0	1	2	3
s_p	-0.65	-0.6437583	-0.6437624	-0.6437624

Exercise 3.2.5 Find the solution of $y''=2(1+x)^{-1}y^2$ subject to $y(0)=1$, $y(1)={}^1/_2$, using the simple shooting method in which the estimate at each stage is improved using Newton's method.

The need to find F'(s) when using Newton's method makes this algorithm somewhat counterproductive for relatively simple problems and it is perhaps easier to replace the iteration given by equation (3.2.7) by the Secant Method

$$s_{p+1}=s_p - F(s_p)\ (s_p-s_{p-1})/[F(s_p)-F(s_{p-1})]. \qquad (3.2.8)$$

(For an explanation of the Secant Method see Exercise 1.3.3.)

Example 3.2.5 Compute a solution for $y''=2(1+x)^{-1}y^2$, subject to $y(0)=1$, $y(1)=1$, using the simple shooting method combined with the secant iteration (3.2.8).

Solution. Using the Classical Runge-Kutta method with h=0.01 the computed approximation for $y(1)=y(100\times h)$ is given by Y_{100}. Taking $z(0)=-0.65$ produces $Y_{100}=0.9909067$ whilst $z(0)=-0.64$ gives $Y_{100}=1.0054916$. Now define $F(s)=Y_N(s)-y(1)=Y_N(s)-1$ then $F(-0.65)=0.9909075-1=0.0090925$ and $F(-0.64)=0.0054916$. From equation (3.2.8) the next estimate for $z(0)$ is s_2 and is given by

$$s_2=-0.64 - {}^{[0.0054916][(-0.64) - (-0.65)]}/_{[1.0054916-0.9909075]}$$
$$=-0.643770731.$$

Using this value for $z(0)$ produces $Y_{100}=0.999987759$, which in turn gives $s_3=-0.643762345$ and $Y_{100}=0.999999992$. This produces $s_4=-0.643762339$ which gives $Y_N=1.00000000$.

Exercise 3.2.2 (revisited) Compute a solution for $y''=2(1+x)^{-1}y^2$, subject to $y(0)=1, y(1)={}^1/_2$, using a simple shooting method which combines the Classical Runge-Kutta Method and the secant iteration (3.2.8).

3.3 Multiple Shooting Methods

The simplicity of the simple shooting method is very attractive; however it can suffer from induced instability if we attempt to compute a bounded solution for an equation which has a general solution which grows rapidly. This was demonstrated in Example 3.2.2 where the general solution was $y(x)=Ae^{30x}+Be^{-30x}+1-2x$ and only the choice $y(0)=1$, $y'(0)=-2$ gives both A and B as zero. This problem can be controlled to some extent by using the multiple or parallel shooting method proposed by Osborne (1969). We illustrate this method with an example.

Example 3.3.1 Compute a solution of the boundary value problem

$$y''=30^2(y-1+2x), \text{ subject to } y(0)=1 \text{ and } y(1)=-1.$$

Solution. As with the simple shooting method we estimate any unknown values at x=0 and turn the problem into one of initial value type. For the given problem $y'(0)$ is unknown and so we set $y'(0)=d_0$ and compute a solution using any of the methods discussed in the previous chapter. Unless $d_0=-2$, (the correct value), the numerical solution will grow rapidly. Let us suppose that when x=X the solution is in danger of growing too large, in which case we restart the numerical integration of the differential equation estimating new values for both $y(X)$ and $y'(X)$, e.g. we set $y'(X)=f_1$ and $y'(X)=d_1$ and then compute a solution of the differential equation for values of x>X subject to these initial conditions. If for some value of x>X the solution again grows too large we can restart again but for the sake of simplicity let us suppose that we are able to find a reasonable approximation for the solution at x=1. Next we compare this computed value with the boundary condition $y(1)=-1$. If in the unlikely event that it agrees and the computed solution is continuous at x=X then we have found the required solution. On the other hand, if this is not the case then we must adjust the estimated parameters d_0,f_1,d_1 so that the boundary condition at x=1 satisfied and also ensure that the computed solution is continuous at x=X. This is equivalent to solving the following equations:-

$$F_1(d_0,f_1,d_1)=y_0(X) - f_1 = 0, \text{ (continuity of } y \text{ at } x=X), \quad (3.3.1)$$
$$F_2(d_0,f_1,d_1)=y'_0(X) - d_1 = 0, \text{ (continuity of } y' \text{ at } x=X), \quad (3.3.2)$$
$$F_3(d_0,f_1,d_1)=y_1(1) - y(1) = 0, \quad (3.3.3)$$
(to satisfy the boundary condition at x=1),

where $y_0(x)$ is the solution of the differential equation on the interval (0,X) and y_1 the solution on (X,1). This gives a system of 3 equations in the 3 unknowns which are solved by a Newton Iteration to obtain d_0, f_1 and d_1. Let us suppose that d_0^*, f_1^*, d_1^* are close to d_0,f_1 and d_1, respectively, i.e. $d_0=d_0^*+\varepsilon_0$, $f_1=f_1^*+\varepsilon_1$ and

$d_1=d_1^*+\varepsilon_2$, then

$$F_i(d_0^*+\varepsilon_1, f_1^*+\varepsilon_1, d_1^*+\varepsilon_2)=0, \qquad i=1,2,3,$$

which may be expanded to give

$$0 \simeq F_i(d_0, f_1, d_1) + \varepsilon_0 \partial F_i/\partial d_0 + \varepsilon_1 \partial F_i/\partial f_1 + \varepsilon_2 \partial F_i/\partial d_1 + O(\varepsilon_i^2),$$

$i=1,2,3$. If we ignore the terms $O(\varepsilon_i^2)$ and above, this gives a system of 3 linear equations to find ε_0, ε_1 and ε_2. These corrections are then added to d_0^*, f_1^* and d_1^* to define an improved solution and the process is repeated until "convergence" is achieved. We could replace the partial derivatives $\partial F_i/\partial d_0$, $\partial F_i/\partial f_1$, and $\partial F_i/\partial d_1$, $i=1,2,3$, by suitable difference approximations, as was done to produce the secant method (3.2.8) from the iteration (3.2.7). Alternatively, we might decide to update them only when the solution is changing rapidly. The extension to the case when more than one restart is required is obvious and though the growth in the number of non-linear equations which need to be solved is unattractive the multiple shooting method can be very effective. Further details are given by Deuflhard (1980), Stoer and Burlirsch (1979) and Hall and Watt (1976).

For Example 3.3.1 we take $y'(0)=d_0^*=-1.5$ and apply the Classical Runge-Kutta method, with $h=0.01$, which produces a difference approximation for $y_0(1/2)$ and $y'_0(1/2)$ given by 27220.3189 and 816607.566. We restart the integration at $x=1/2$ estimating both $y(1/2)$ and $y'(1/2)$ as $y_1(1/2)=f_1^*=0$ and $y'(1/2)=y'_1(1/2)=d_1^*=-1$ which produces $y_1(1)=54439.6379$. Equations (3.3.1), (3.3.2) and (3.3.3) then give the values:- $F_1(-1.5,0,-1)=27220.3189$, $F_2(-1.5,0,-1)=816608.566$ and $F_3(-1.5,0,-1)=54440.6379$ from which we obtain the correction $\varepsilon_0=-0.4999994$, $\varepsilon_1=0.000000$, $\varepsilon_2=-0.99999$ which added to d_0^*, f_1^* and d_1^* give new estimates for $d_0 \simeq -1.999999$, $f_1 \simeq 0.0$ and $d_1 \simeq -1.99999$. A second iteration gives $d_0=-2$, $f_1=0$ and $d_1=-2$. The rapid convergence in this example is due to the linearity of the differential equation but provided a good first estimate is supplied, convergence is achieved quickly.

3.4 Finite Difference Methods

We consider now an alternative to the shooting method which does not suffer from the problem of induced instability. We approximate all the derivatives in the equation $y''=f(x,y,y')$ and effectively reduce the problem to solving a system of linear equations.

We begin by considering the linear two point boundary value problem

$$y'' = p(x)y + q(x), \quad y(a)=\alpha, \quad y(b)=\beta, \tag{3.4.1}$$

and at each point $x_n=a+nh$, $n=1,2,\ldots,N-1$, $Nh=(b-a)$, we replace $y''(x)$ by a suitable difference approximation. For example, using the centred difference approximation (2.12.2) produces

$$\frac{y(x+h)-2y(x)+y(x-h)}{h^2} = p(x)y(x) + q(x) + E(h), \tag{3.4.2}$$

where the local discretization error $E(h)$ is $^{h^2}/_{12}\, ^{d^4y}/_{dx^4}(\theta)$, $x-h\leqslant\theta\leqslant x+h$. If we approximate $y(x_n)$ by Y_n and then select h so that the local discretization error is "negligible" equation (3.4.2) becomes

$$Y_{n+1}-2Y_n+Y_{n-1} = h^2[\ p(x_n)Y_n + q(x_n)]. \tag{3.4.3}$$

For n=1 this gives

$$Y_2-2Y_1+Y_0 = h^2[p(x_1)Y_1 + q(x_1)],$$

for n=2

$$Y_3-2Y_2+Y_1 = h^2[p(x_2)Y_2 + q(x_2)],$$

and so on until $n=N-1$ when

$$Y_N-2Y_{N-1}+Y_{N-2} = h^2[p(x_{N-1})Y_{N-1} + q(x_{N-1})].$$

But $Y_0=y(x_0)=y(a)=\alpha$ and $Y_N=y(x_N)=y(b)=\beta$ in which case the equations (3.4.3) give a system of N-1 linear equations in the N-1 unknowns

$Y_1, Y_2, \ldots, Y_{N-1}$, i.e.

$$\mathbf{AY} = \mathbf{b}, \tag{3.4.4}$$

where

$$\mathbf{A} = \begin{bmatrix} -(2+h^2p(x_1)) & 1 & & & \\ 1 & -(2+h^2p(x_2)) & 1 & & \\ & \cdot & \cdot & \cdot & \\ & & \cdot & \cdot & \cdot \\ & & & \cdot & \cdot & \cdot \\ & & & & 1 & -(2+h^2p(x_{N-1})) \end{bmatrix}$$

$$\mathbf{Y} = \begin{bmatrix} Y_1 \\ Y_2 \\ \cdot \\ \cdot \\ \cdot \\ Y_{N-1} \end{bmatrix}, \qquad \mathbf{b} = \begin{bmatrix} h^2q(x_1) - \alpha \\ h^2q(x_2) \\ \cdot \\ \cdot \\ \cdot \\ h^2q(x_{N-1}) - \beta \end{bmatrix}.$$

The system of equations (3.4.4) has a unique solution, **Y**, provided A is non-singular. If there exists a positive constant P such that $p(x) \geqslant P > 0$, for all x in (a,b), then A is strictly diagonally dominant, therefore, it can have no zero eigenvalues and is non-singular as required. (See Appendix 1.) Furthermore, the equations are tridiagonal and may be solved rapidly and economically using the LU decomposition given in Chapter 1.

We can examine the errors in this process by subtracting equation (3.4.3) from $h^2 \times$(3.4.2). Writing $e_n = y(x_n) - Y_n$ this gives

$$e_{n+1} - 2e_n + e_{n-1} = h^2p(x_n)e_n + h^2E_h, \quad n=1,2\ldots,N-1. \tag{3.4.5}$$

As there are only a finite number of e_n's there must be one which is at least as large as any other, let this be the mth, i.e. $|e_m| \geqslant |e_n|$, $n=1,2,\ldots,N-1$. Now consider equation (3.4.5) when n=m,

$$(2 + h^2 p(x_m))e_m = e_{m+1} + e_{m-1} - h^2 E_h.$$

Taking moduli we have

$$(2 + h^2 p(x_m))|e_m| \leqslant |e_{m+1}| + |e_{m-1}| + h^2|E| \leqslant |e_m| + |e_m| + h^2 E,$$

where $E = \max|E(h)| = h^2 \sup|d^4y/dx^4|/12$, therefore,

$$h^2 p(x_m)|e_m| \leqslant h^2 E.$$

Since we have assumed that $p(x) \geqslant P > 0$ we obtain

$$\max_{n=1,\ldots,N-1} |e_n| = |e_m| \leqslant E/P \leqslant h^2/12 \sup|y^{(iv)}|, \qquad (3.4.6)$$

which tends to zero as $h \to 0$ provided sufficient derivatives of y exist. We conclude that the difference solution, Y_n, $n=1,2,\ldots,N-1$, converges to the solution of the differential equation as $h \to 0$.

The boundary conditions given for equation (3.4.1) are not the most general possible; we may have to satisfy the differential equation together with the boundary conditions

$$\begin{aligned} \alpha_1 y(a) + \alpha_2 y'(a) &= \alpha_3, \\ \beta_1 y(b) + \beta_2 y'(b) &= \beta_3, \end{aligned} \qquad (3.4.7)$$

or even the general separable boundary conditions $G_1(y(a),y'(a))=0$ and $G_2(y(b),y'(b))=0$. If α_2, (or β_2), is non-zero then it is necessary to replace $y'(a)$ by a difference approximation, for example, $y'(a) \simeq (Y_1 - Y_0)/h$. Such an approximation for $y'(a)$ has an error $O(h)$ which will detract from the overall $O(h^2)$ accuracy of the difference method (3.4.3). This problem can be overcome by introducing a fictitious or spurious point at $x = x_{-1} = a-h$ at which there is a value denoted by Y_{-1}. Alternatively, the centred difference approximation $y'(a) = (y(a+h) - y(a-h))/2h$ has an error

$O(h^2)$, and we replace the first equation of (3.4.7) by

$$\alpha_1 Y_o + \alpha_2 (Y_1 - Y_{-1})/2h = \alpha_3, \qquad (3.4.8)$$

which is of the same order of accuracy as the difference equation (3.4.3). In addition we approximate the differential equation at $x=x_o=a$ to give

$$Y_1 - 2Y_o + Y_{-1} = h^2[p(a)Y_o + q(a)]. \qquad (3.4.9)$$

We then eliminate the spurious value Y_{-1} between equations (3.4.8) and (3.4.9) and the resulting equation is combined with (3.4.4). If necessary a similar procedure can be carried out at x=b.

Example 3.4.1 Find a numerical solution for the two point boundary value problem

$$(1+x)^2 y'' = 2y - 4,$$

subject to $y(0)=0$ and $y(1)-2y'(1)=0$.

Solution. We divide the interval (0,1) into 4 sub-intervals each of width 0.25 as shown in Figure 3.4.1 and let Y_n be a computed approximation for $y(nh)$, n=0,1,2,3,4. At x=0 we have $Y_o=y(0)=0$.

Figure 3.4.1

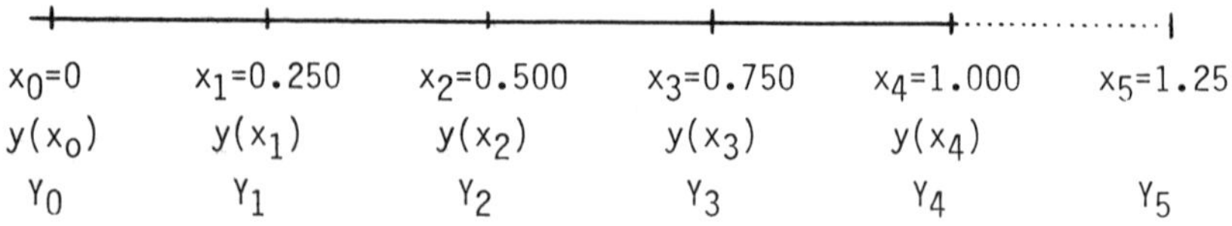

As we have a derivative boundary condition at x=1 we introduce the spurious point x_5 associated with which is the value Y_5. Firstly, we apply equation (3.4.3) at $x_1=0.25$ to give

$$Y_2-2Y_1+Y_0 = 0.25^2\,(2Y_1-4)/(1+0.25)^2\ ,$$

at $x_2=0.5$:

$$Y_3-2Y_2+Y_1 = 0.25^2\,(2Y_2-4)/(1+0.5)^2\ ,$$

and at $x_3=0.75$:

$$Y_4-2Y_3+Y_2 = 0.25^2\,(2Y_3-4)/(1+0.75)^2\ .$$

At x_4 we have two equations, the first comes from the difference method, i.e.

$$Y_5-2Y_4+Y_3 = 0.25^2\,(2Y_4-4)/2^2$$

and the second from approximating the given boundary condition as in (3.4.8)

$$Y_4 - 2(Y_5-Y_3)/(2\times 0.25) = 0,\ \text{i.e.}\ \ Y_4 - 4Y_5 + 4Y_3 = 0.$$

We next eliminate Y_5 to give

$$1/4\times(Y_4 + 4Y_3) - 2Y_4 + Y_3 = 0.25^2\,(2Y_4-4)/4.$$

Collecting together these equations gives

$$\begin{bmatrix} -2.08 & 1 & 0 & 0 \\ 1 & -2.0556 & 1 & 0 \\ 0 & 1 & -2.0408 & 1 \\ 0 & 0 & 2 & -1.7813 \end{bmatrix} \begin{bmatrix} Y_1 \\ Y_2 \\ Y_3 \\ Y_4 \end{bmatrix} = \begin{bmatrix} -0.1600 \\ -0.1111 \\ -0.0816 \\ -0.0625 \end{bmatrix}$$

Unfortunately, because of the derivative boundary condition this is not a diagonally dominant system; however, using partial pivoting and working to 5 decimal places we obtain $Y_1=0.38904$, $Y_2=0.64922$, $Y_3=0.83435$ and $Y_4=0.97190$ which can be compared with the exact values which are 0.4, 0.6667, 0.85714 and 1 respectively.

Exercise 3.4.1 Compute an approximate solution of $y''=y$, subject to $y(0)=0$ and $y(1)=1$, using the finite difference equation (3.4.3) with $h=0.25$. Compare the computed approximation with the solution of the differential equation which satisfies these boundary conditions.

Exercise 3.4.2 Compute an approximate solution of $y''=y$, subject to $y(0)=0$ and $y(1)=1$, using Numerov's method given in Exercise 2.12.1 with $h=0.25$. Compare the computed result with that obtained in Exercise 3.4.1.

For the more general non-linear two point boundary value problem

$$y''=f(x,y,y'), \qquad y(a)=\alpha,\ y(b)=\beta,$$

we again seek an approximate solution which is defined at the points $x_n=a+nh$, $n=1,2,\ldots,N-1$, $Nh=(b-a)$ and is denoted by Y_n. The simplest approximation for this equation is given by

$$Y_{n+1}-2Y_n+Y_{n-1} = h^2 f(x_n, Y_n, (Y_{n+1}-Y_{n-1})/2h), \qquad (3.4.10)$$

for $n=1,2,\ldots,N-1$, with $Y_0=\alpha$ and $Y_N=\beta$. This gives a system of N-1 non-linear equations

$$A\,Y = h^2\, f(x,Y),$$

where

$$A = \begin{bmatrix} -2 & 1 & 0 & & 0 \\ 1 & -2 & 1 & & 0 \\ & \cdot & \cdot & \cdot & \\ & & \cdot & \cdot & \cdot \\ & & & 1 & -2 \end{bmatrix}, \quad Y = \begin{bmatrix} Y_1 \\ Y_2 \\ \cdot \\ \cdot \\ Y_{N-1} \end{bmatrix}, \quad f = \begin{bmatrix} f(x_1, Y_1, (Y_2-Y_0)/2h)-\alpha \\ f(x_2, Y_2, (Y_3-Y_1)/2h) \\ \cdot \\ \cdot \\ f(x_{N-1}, Y_{N-1}, (Y_N-Y_{N-2})/2h)-\beta \end{bmatrix}$$

These equations can be solved using the simple iteration

$$A\,Y^{r+1} = h^2 f(x, Y^r). \quad r=1,2,.. \qquad (3.4.11)$$

If we assume that f satisfies a Lipschitz condition with respect to $\mathbf{Y}$ then

$$A(\mathbf{Y}^{r+1} - \mathbf{Y}^{r}) = h^2 \left(\mathbf{f}(x,\mathbf{Y}^r)-\mathbf{f}(x,\mathbf{Y}^{r-1})\right)$$

gives

$$\|\mathbf{Y}^{r+1}-\mathbf{Y}^{r}\| \leqslant h^2 \|A^{-1}\| \; L \; \|\mathbf{Y}^{r}-\mathbf{Y}^{r-1}\|,$$

therefore, the iteration (3.4.11) will converge provided $Lh^2\|A^{-1}\|<1$, i.e. if h is sufficiently small. Since A is a symmetric matrix $\|A^{-1}\|_2 = |\text{maximum eigenvalue of } A^{-1}|=[\text{minimum eigenvalue of } A]^{-1}$. The eigenvalues of A are given by $\lambda_j=-2+2\cos{}^{j\pi}/_{(N-1)}$, $j=1,\ldots,N-1$, and so $\|A^{-1}\|={}^1/(2-2\cos\pi/(N-1))\simeq[{}^{(N-1)}/_{\pi}]^2$. Therefore, the iteration (3.4.11) will converge if $({}^{(N-1)h}/_{\pi})^2L<1$. The relatively slow rate at which this iteration converges means that higher order methods are to be preferred.

Returning to the equation (3.4.10) and writing them in matrix form we have that

$$A\mathbf{Y} = h^2\mathbf{f}(x,\mathbf{Y})$$

More generally, we seek a vector Y such that the function given by $F(\mathbf{Y})=A\mathbf{Y}-h^2f(x,\mathbf{Y})$ vanishes. Let us suppose that Y^r is an approximation for $\mathbf{Y}$ and that $\mathbf{Y}=\mathbf{Y}^r+\boldsymbol{\varepsilon}^r$ then

$$\mathbf{0} = F(\mathbf{Y}) = F(\mathbf{Y}^r+\boldsymbol{\varepsilon}^r) = F(\mathbf{Y}^r)+F'(\mathbf{Y}^r)\boldsymbol{\varepsilon}^r+ \ldots.$$

where F' denotes the matrix of first partial derivatives of the vector function f and is called its Jacobian. If $\boldsymbol{\varepsilon}^r$ is "small" then $F'(\mathbf{Y}^r)\boldsymbol{\varepsilon}^r=-F(\mathbf{Y}^r)$ which may be solved for $\boldsymbol{\varepsilon}^r$ and hence we define $\mathbf{Y}^{r+1}=\mathbf{Y}^r+\boldsymbol{\varepsilon}^r$ which should give a better estimate for $\mathbf{Y}$. Notice that if F is a scalar function this method will reduce to the usual Newton iteration. When this technique is applied to boundary value problems the Jacobian is tridiagonal and so its application is relatively straightforward. In some algorithms the Jacobian is either estimated by finding a numerical approximation for the partial derivatives of F or else it is only recomputed when the iterates change rapidly. Further details of the application of this method are given by Stoer and Bulirsch (1980) and Hall and Watt (1976). An analysis of the solution of non-linear systems of equations is given by Kantorovich and Akilov (1982).

Example 3.4.2 Find a solution of $y''=[^2/_{(1+x)}]y^2$, $y(0)=1$, $y(1)=0.5$.

Solution. As y' does not appear we could use the $O(h^4)$ Numerov's method to approximate the differential equation, but for simplicity we use the simple $O(h^2)$ approximation (3.4.10). We divide the interval (0,1) into N sub-intervals each of width h and at each point x=nh we compute an approximation for the solution $y(x_n)$ which will be denoted by Y_n. The difference solution Y_n, n=1,..,N-1, then satisfies

$$Y_{n+1}-2Y_n+Y_{n-1} = 2h^2\, Y_n^2/(1+nh), \quad n=1,2,\ldots,N-1. \quad (3.4.12)$$

Let $\{Y_n\}^0$ be an approximation for Y_n and set $Y_n=\{Y_n\}^0+\{\varepsilon_n\}^0$ then

$$(\{Y_{n+1}\}^0+\{\varepsilon_{n+1}\}^0) - 2(\{Y_n\}^0+\{\varepsilon_n\}^0) + (\{Y_{n-1}\}^0+\{\varepsilon_{n-1}\}^0)$$

$$= {}^{2h^2}/_{(1+nh)}\, [\{Y_n\}^0 + \{\varepsilon_n\}^0]^2, \qquad n=1,2,\ldots,N-1.$$

If $\{\varepsilon_n\}^0$, n=1,2,...,N-1, are so small that we can ignore terms like $(\{\varepsilon_n\}^0)^2$, then

$$\{\varepsilon_{n+1}\}^0 - (2 + {}^{4h^2}/_{(1+nh)}\{Y_n\}^0)\{\varepsilon_n\}^0 + \{\varepsilon_{n-1}\}^0$$

$$= -(\{Y_{n+1}\}^0-2\{Y_n\}^0+\{Y_{n-1}\}^0 - {}^{2h^2}/_{(1+nh)}[\{Y_n\}^0]^2), \qquad n=1,2,\ldots,N-1,$$

with $\{\varepsilon_0\}^0=\{\varepsilon_N\}^0=0$. This set of linear equations is solved to find $\{\varepsilon_n\}^0$, n=1,2,...,N-1, which are then added to $\{Y_n\}^0$, n=1,..,N-1, to give an improved estimate $\{Y_n\}^1$, n=1,2,...,N-1, and the process is iterated to convergence. Taking N=10 gives the results which are given in Table 3.4.1.

Table 3.4.1 Approximate solutions of $(1+x)y''=2y^2$, $y(0)=1$, $y(1)=0.5$ obtained by solving equation (3.4.12)

Iteration	x=0.2	x=0.4	x=0.6	x=0.8
0	0.900000	0.800000	0.700000	0.600000
1	0.833181	0.714200	0.624419	0.554067
2	0.832226	0.713110	0.623504	0.553566
:	:	:	:	:
5	0.832217	0.713104	0.623500	0.553563
6	0.832217	0.713104	0.623500	0.553563
y(x)	0.833333	0.714286	0.625000	0.555556
error	0.001163	0.001182	0.001500	0.001993

(The initial estimate $\{Y_n\}^0$ is put equal to the straight line joining the given boundary conditions. The iteration is terminated when the maximum difference between successive estimates is less that 10^{-6}.)

Although the results given in Table 3.4.1 are rather crude we will give details later of how to refine such a solution. Finally we give a global bound on $e_n=|y(x_n)-Y_n|$ in the special case of the equation $y''=f(x,y)$, subject to $y(a)=\alpha$ and $y(b)=\beta$.

Theorem 3.4.1 Consider the two point boundary value problem $y''=f(x,y)$, subject to $y(a)=\alpha$ and $y(b)=\beta$, which is to be solved using the difference method

$$\frac{Y_{n+1}-2Y_n+Y_{n-1}}{h^2} = f(x_n,Y_n), \qquad n=1,2,\ldots,N-1,$$

with $Y_0=\alpha$ and $Y_N=\beta$. We assume that $\sup|d^4y/dx^4|$ exists and is equal to M_4 and that $f_y\geq 0$ for $a\leq x\leq b$ and $-\infty < y < \infty$, then the global error satisfies

$$|y(x_n) - Y_n| \leq M_4h^2/24\ (x_n-a)(b-x_n). \qquad (3.4.13)$$

Proof. We have that

$$y(x_{n+1})-2y(x_n)+y(x_{n-1})=h^2f(x_n,y(x_n))+h^4/_{12}y^{iv}(\theta_n), \quad (3.4.14)$$
$$n=1,2,\ldots,N-1,$$

where θ_n lies between x_{n-1} and x_{n+1}, and

$$Y_{n+1}-2Y_n+Y_{n-1} = h^2f(x_n,Y_n), \quad n=1,2,\ldots,N-1. \quad (3.4.15)$$

Let $e_n=y(x_n)-Y_n$ and then subtracting (3.4.15) from (3.4.14) we have that

$$e_{n+1}-2e_n+e_{n-1}=h^2e_nf_y(x_n,\phi_n)+h^4/_{12}y^{iv}(\theta_n), \quad (3.4.16)$$

where ϕ_n lies between $y(x_n)$ and Y_n. Now consider the difference equation

$$u_{n+1}-b_nu_n+u_{n-1}=c_n, \quad n=1,2,\ldots,N-1, \quad (3.4.17)$$

where $b_n\geqslant 2$ and $c_n\geqslant 0$, with u_o and $u_N \leqslant 0$. We have that

$$u_n = \frac{u_{n+1}+u_{n-1}}{b_n} - \frac{c_n}{b_n} \leqslant \frac{u_{n+1}+u_{n-1}}{b_n}$$

and so u_n cannot be greater than both u_{n+1} and u_{n-1}. Therefore, the maximum value of any u_n must either be when n=0 or n=N, i.e. either $u_n\leqslant u_o\leqslant 0$ or $u_n\leqslant u_N\leqslant 0$. Now we examine

$$z_{n+1}-2z_n+z_{n-1}= h^4M_4/_{12}, \quad (3.4.18)$$

subject to the boundary conditions $z_o=z_N=0$. This recurrence relation has the solution $z_n=-h^2M_4/_{24}\ (x_n-a)(b-x_n)$. Adding (3.4.18) to (3.4.16) and then re-arranging the result gives

$$(z_{n+1}+e_{n+1})-(2+h^2f_y(x_n,\theta_n)(z_n+e_n)+(z_{n-1}+e_{n-1})$$

$$= h^4(M_4+y^{iv}(\phi_n))/_{12}-h^2f_y(x_n,\theta_n)z_n. \quad (3.4.19)$$

The right hand side of (3.4.19) is non-negative since $z_n \leq 0$ and $f_y \geq 0$. Furthermore, $z_0 = e_0 = 0$ and $z_N = e_N = 0$ which means that (3.4.19) is of the form (3.4.17) from which we conclude that $z_n + e_n \leq 0$. Likewise, subtracting (3.4.18) from (3.4.16) gives that $z_n - e_n \leq 0$ and so $|e_n| < -z_n = {}^{h^2M_4}/_{24}\,(x_n - a)(b - x_n)$ as required.

The error bound (3.4.13) is usually very much larger than the actual error and requires some knowledge of ${}^{d^4y}/_{dx^4}$ but it does show that the global error is $O(h^2)$ and that the method is convergent of order 2. We have assumed that it has been possible to find Y_n exactly though in general this will not be the case and in addition there will be inaccuracies due to the accumulation of rounding errors. We can amend the proof of the Theorem 3.4.1 as follows. Let us suppose that instead of solving equation (3.4.15) exactly we actually solve

$$Z_{n+1} - 2Z_n + Z_{n-1} = h^2 f(x_n, Z_n) + r_n, \qquad n=1,2,\ldots,N-1,$$

where r_n is the error due to solving a set of simultaneous equations. If we assume that there exists a constant r and an integer k such that $|r_n| < rh^k$, $n=1,2,\ldots,N-1$, then provided $k \geq 4$ we need only add a term rh^{k-4} to ${}^1/_{12}M_4$ in the error bound (3.4.13).

3.4.1 Improvement of the computed solution

Using low order differences to replace any derivative means that the solution will itself be of relatively low order and so it will be necessary to refine it in some way. There are basically two different approaches to tackle this problem. The first is based on the idea of h^2-extrapolation which has been discussed previously in connection with numerical integration in Section 1.5 and in connection with initial value problems in Section 2.9. The second method, which is called difference correction, attempts to take into account terms in the discretization error which were assumed to be negligible.

Theorem 3.4.1 demonstrates that the finite difference method given by equation (3.4.15) is convergent of order 2 provided sufficient derivatives of the solution exist. This result means that if at some point x the solution is y(x) for which we have computed an approximation Y(x,h) using a step size h then

$$y(x) - Y(x,h) = Ah^2 + \text{Higher Order Terms.} \quad (3.4.20)$$

Similarly given $Y(x,{}^h/_2)$ and $Y(x,{}^h/_4)$ we can eliminate y(x) and A to give

$$(Y(x,h/2)-Y(x,h))/(Y(x,h/4)-Y(x,h/2)) \simeq 4.$$

If this can be shown to be the case then we can have some confidence that sufficient derivatives of y exist for (3.4.20) to be valid.

Example 3.4.3 Compute a solution for $(1+x)^2y''=2y-4$, subject to $y(0)=0$, $y(1)-2y'(1)=0$ using values of h=0.25, 0.125 and 0.0625, and verify that (3.4.20) is satisfied.

Solution. The solution of this equation with h=0.25 is given by Example 3.4.1. Repeating the calculation with h=0.125 and 0.0625 gives the results in Table 3.4.2.

Table 3.4.2. The solution of $(1+x)^2y''=2y-4$, $y(0)=0$, $y(1)-2y'(1)=0$ for h=0.25, 0.125, 0.0625.

	h=0.25 (1)	h=0.125 (2)	h=0.0625 (3)	(2)-(1)	(3)-(2)	Ratio	Extrapolated
x=0.25	0.389049	0.397164	0.399284	0.00812	0.00212	3.83	0.399991
x=0.50	0.649221	0.662173	0.665534	0.01295	0.00361	3.85	0.666665
x=0.75	0.854351	0.851290	0.855669	0.01694	0.00437	3.87	0.857127
x=1.00	0.971903	0.992798	0.998188	0.02090	0.00539	3.87	0.999985

As the ratio of successive differences is approximately 4 we may have some confidence in the existence of sufficient derivatives for equation (3.4.20) to be valid. Using this equation and the corresponding equation for $h/2$, i.e.

$$y(x) - Y(x,h/2) = A(h/2)^2 + ...$$

and eliminating A gives $(4Y(x,h/2)-Y(x,h))/3$ as an $O(h^4)$ approximation for $y(x)$. This gives the results in the column labelled "Extrapolated" which may be compared with the solution of the differential equation satisfying the given boundary conditions, i.e. $y(x)=2x/(1+x)$. In the same way we can check for the existence of higher order terms in equation (3.4.20) and eliminate them as far as is necessary. This process is called Deferred Correction.

The validity of equation (3.4.20) is based upon the assumption that the solution has sufficient derivatives, a fact which is not always evident. This type of extrapolation must be carried out with care as was demonstrated by Fox (1962). For example, the function $y(x)=1/2x^2\log(x)$ satisfies the differential equation

$$xy'' - y' - x = 0, \text{ subject to } y(0)=y(1)=0,$$

but numerical results suggest that h^2-extrapolation applied to the results obtained using (3.4.10) is not valid. It can be shown that

$$y(x) - Y(x,h) = - h^2/8\,[(x^2-1)\log(h) + \log(x)] + ...$$

(see Mayers (1964)) which explains why the simple h^2-extrapolation fails. The presence of the log term in the solution is not the cause of the problem for the same function satisfies the equation

$$x^2y'' - 2y - 3/2\, x^2 = 0, \text{ subject to } y(0)=y(1)=0,$$

and yet simple extrapolation works well. It can be shown that if the difference method (3.4.10) is applied to this equation then

$$y(x) - Y(x,h) = {}^{h^2}/_{24}(1-x^2) + {}^{h^4}/_{240}(x^2-{}^1/_{x^2}) + ..$$

Furthermore, even if the solution of the differential equation has all the derivatives we could desire there is still no guarantee that the simple h^2-extrapolation will be sufficient. For example, the function $y(x)={}^{x^3}/_9$ satisfies the differential equation

$$xy'' + y' = x^2, \text{ subject to } y(0)=0,\ y(1)={}^1/_9,$$

yet

$$y(x) - Y(x,h) = {}^{h^2}/_9(x-1) + {}^1/_9\ {}^{h^2}/_{\log(h)} \log(x) + ...$$

and though it is possible to eliminate both these terms the simple h^2-extrapolation will fail. The moral is clear: this type of extrapolation must be used with care and sufficient analysis.

Exercise 3.4.3 Compute an approximate solution for $y''=y$, $y(0)=0$, $y(1)=1$, using (3.4.10) with $h=0.25$, 0.125 and 0.0625 and investigate whether (3.4.20) is valid. Extrapolate the computed solutions and compare them with the particular solution of the differential equation which satisfies the given boundary conditions.

As an alternative to the extrapolation process described above we will now consider the difference correction procedure proposed by Fox (1962). The equation $y''=f(x,y,y')$ can be approximated by

$$y(x_{n+1})-2y(x_n)+y(x_{n-1})=h^2f(x_n,y(x_n),{}^{[y(x_{n+1})-y(x_{n-1})]}/_{2h}) + h^2E(h)$$
$$n=1,2,...,N-1,$$

where $E(h)={}^{h^2}/_{12}\ {}^{d^4y}/_{dx^4} + O(h^4)$, which may be written as the system of equations

$$\mathbf{Ay} = h^2\mathbf{f}(x,\mathbf{y}) + E(h), \qquad (3.4.21)$$

where A and f are as for (3.4.11) and $y=(y(x_1),y(x_2),..,y(x_{N-1}))^T$. Previously we selected h sufficiently small that E(h) was negligible and then found an approximate solution Y such that $\mathbf{AY}=h^2\mathbf{f}(x,\mathbf{Y})$. We now extend this procedure as follows. Having found Y we use it to estimate $E(h)=E(h,Y)$ and then use the iteration

$$AY^{r+1} = h^2 f(x,Y^r) + E(h,Y^r) \qquad (3.4.22)$$

to improve successively the computed solution by estimating any higher order derivatives of y, i.e. y^{iv}, y^{vi},.... For example we use $y^{iv}(x) \simeq (y(x-2h)-4y(x-h)+6y(x)-4y(x+h)+y(x+2h))/_{h^4}$ and similarly for $y^{(vi)}$, ... This may mean computing values outside the interval where the solution is required but can be very worthwhile as demonstrated by the following example.

Example 3.4.4 Use the process of difference correction to find a solution of $(1+x)^2y''=2y-4$, $y(0)=0$, $y(1)-2y'(1)=0$.

Solution. Taking N=10 we compute a solution using (3.4.21) neglecting the local discretization error. The resulting approximation is denoted by $\{Y_n\}^0$, n=1,2,..,10. We then apply the difference correction algorithm until the entries of $E(h,Y^r)$ are less than 1 part in 10^7. The results obtained are given in Table 3.4.3.

Table 3.4.3. The difference correction algorithm applied to $(1+x)^2y''=2y-4$, $y(0)=0$, $y(1)-2y'(1)=0$.

x	$\{Y_n\}^0$	$\{Y_n\}^3$	Exact Value
0.1	0.18092462	0.18181813	0.1818182
0.2	0.33178190	0.33333331	0.3333333
0.3	0.45946950	0.46153843	0.4615384
0.4	0.56892598	0.57142854	0.5714285
0.5	0.66377967	0.66666662	0.6666667
0.6	0.74675585	0.75000000	0.7500000
0.7	0.81994107	0.82352936	0.8235294
0.8	0.88495976	0.88888883	0.8888889
0.9	0.94309555	0.94736838	0.9473684
1.0	0.99537588	1.00000000	1.0000000

The particular solution of the differential equation satisfying the given boundary conditions is $y(x)={}^{2x}/(1+x)$ and the value of such a refinement of the solution is clear.

Example 3.4.5 Compare the extrapolation and difference correction methods for the non-linear 2 point boundary value problem $(1+x)y''=2y^2$, $y(0)=1$, $y(1)=0.5$.

Solution. Using (3.4.10) we compute solutions for the differential equation with N=4,8,16 and compare the calculated results.

Table 3.4.4 An examination of the computed solution for $(1+x)y''=2y^2$, $y(0)=1$, $y(1)=0.5$.

x	h=0.25 (1)	h=0.125 (2)	h=0.0625 (3)	(2)-(1)	(3)-(2)	Ratio
0.25	0.788066	0.797999	0.799693	0.009933	0.001694	5.86
0.50	0.654939	0.664423	0.666233	0.009484	0.001810	5.24
0.75	0.557556	0.568416	0.570752	0.010860	0.002336	4.65

If the error $y(x)-Y_h(x)$ were simply $Ah^2 + ..$ then the entries in the column "ratio" should be approximately 4; from these results we cannot conclude that this is the case. If we were to attempt to use the simple h^2-extrapolation on the entries in columns (2) and (3) we obtain the values 0.8002576, 0.6668367 and 0.5715306 which may be compared with the exact values of the particular solution satisfying the given boundary condition, $y(x)=(1+x)^{-1}$, i.e. 0.8, 0.6666667 and 0.5714285. Clearly there has been no significant improvement based on the assumption that (3.4.20) is valid. It is possible to apply the more general Aitken extrapolation to this problem as follows. If we assume that the principal error is of the form

$$y(x) - Y(x,h) = Ah^p + \ldots.$$

then by computing solutions for h, ${}^h/_2$ and ${}^h/_4$ we can eliminate the term Ah^p. This produces an extrapolated value for y(x) given by

$$Y(x,h,h/_2,h/_4) = \frac{Y(x,h)\ Y(x,h/_4) - Y(x,h/_2)^2}{Y(x,h) - 2Y(x,{}^h/_2) + Y(x,{}^h/_4)}$$

which applied at x=0.5 produces an estimate for the correct value of $y(0.5)={}^2/_3$ of 0.66665999 which is somewhat better. Alternatively, using the difference correction algorithm we obtain the results shown in Table 3.4.5 below.

Table 3.4.5 The Difference Correction Algorithm Applied to $(1+x)y''=2y^2$, $y(0)=1$, $y(1)=0.5$.

x	$\{Y_n\}^0$	$\{Y_n\}^4$	Exact
0.1	0.9078554	0.9090909	0.9090909
0.2	0.8322172	0.8333333	0.8333333
0.3	0.7681221	0.7692307	0.7692307
0.4	0.7131041	0.7142857	0.7142857
0.5	0.6653506	0.6666667	0.6666667
0.6	0.6234996	0.6250000	0.6250000
0.7	0.5805081	0.5882353	0.5882352
0.8	0.5535635	0.5555555	0.5555556
0.9	0.5240237	0.5263157	0.5263157

The correction process is terminated when the entries of $E(h,\mathbf{Y}^r)$ have moduli less than 10^{-7}. Once again we see the value in extrapolation of this type.

Exercise 3.4.4 Compare the simple extrapolation procedure, based on the assumption that the error $y(x)-Y(x,h)$ is $Ah^2 + \ldots.$, with the difference correction algorithm for the equation $y''=y$, $y(0)=0$, $y(1)=1$.

Exercise 3.4.5 Compute an approximate solution of the linear two point boundary value problem $y''=30^2(y-1+2x)$, subject to $y(0)=1$ and $y(1)=-1$ by (i) the simple shooting method and (ii) finite difference methods including extrapolation. Which method is to be preferred for this example (See Example 3.2.2).

3.5 Finite Element Methods

With both the shooting method and finite difference techniques an approximate solution for a given differential equation is computed only at a discrete set of points; if the solution is required at a non-mesh point then it is necessary to interpolate the computed solution. We shall now investigate the **finite element method** which produces a solution which is defined at every point in the interval. For the sake of clarity we begin by considering only linear differential equation which are written in the self adjoint form

$$(p(x)y')' + q(x)y = f(x), \qquad (3.5.1)$$

subject to the homogeneous boundary conditions $y(0)=y(1)=0$. It is relatively easy to show that any linear two point boundary value problem subject to known boundary values can be reduced to this form.

Example 3.5.1 Reduce the equation $(1+x)^2z''=2z-4$, $z(0)=0$, $z(1)=1$ to the self adjoint form (3.5.1).

Solution. The first step is to reduce the boundary conditions to a homogeneous form. This is accomplished by setting $y(x)=z(x)-x$ for then $y(0)=z(0)-0=0$ and $y(1)=z(1)-1=0$ as required. With this change of variable the differential equation becomes $(1+x)^2y''-2y=2x-4$. Differentiating equation (3.5.1) gives $py''+p'y'+qy=f$ and comparing coefficients we see that $p'=0$ which implies that p is constant. Since the given differential equation is linear we may set $p=1$, hence, $q(x)=-2/(1+x)^2$ and $f=(2x-4)/(1+x)^2$, so the given equation becomes

$$(y')' - 2(1+x)^{-2}\, y = (2x-4)/(1+x)^2.$$

The idea of the finite element method is to select a set of linearly independent functions $\phi_j(x)$, $j=1,..,M$, each of which satisfies the conditions $\phi_j(0)=\phi_j(1)=0$ and then to look for an approximate solution of the differential equation of the form

$$Y_M(x) = \sum_{j=1..M} c_j\phi_j(x). \qquad (3.5.2)$$

Notice that $Y_M(x)$ is defined at every point of the interval [0,1] and automatically satisfies the boundary conditions. The problem now is to decide which basis function to use and then how to determine the coefficients c_j, j=1,..,M, in such a way as to make Y_M the best approximation available, i.e. we wish to minimize the error between $y(x)$ and $Y_M(x)$ for values of x between 0 and 1.

Example 3.5.2 Two possible sets of basis functions are

$$\text{(i)}\ \phi_j(x)=\sin j\pi x,\ j=1,\dots,M, \qquad 0\leqslant x\leqslant 1,$$
$$\text{(ii)}\ \phi_j(x)=x(1-x)x^{j-1},\ j=1,\dots,M, \qquad 0\leqslant x\leqslant 1.$$

The problem of determining suitable values for the coefficients c_j may be tackled in three different ways.

3.5.1 The Method of Collocation

Differentiating equation (3.5.2) and substituting the resulting expression into equation (3.5.1) we obtain

$$p(x)\sum c_j\phi''_j(x)+p'(x)\sum c_j\phi'_j(x)+q(x)\sum c_j\phi_j(x)=f(x)$$

or

$$\sum_{j=1..M} [p(x)\phi''_j(x)+p'(x)\phi'_j(x)+q(x)\phi_j(x)]c_j=f(x).$$

If $Y_M(x)$ were the exact solution then this expression would be satisfied identically, but, since we have only M parameters at our disposal we cannot expect it to be satisfied at more than M points x_k, k=1,...,M. The coefficients c_j, j=1,..,M satisfy the linear equations

$$\sum_{j=1..M} [p(x_k)\phi''_j(x_k)+p'(x_k)\phi'_j(x_k) + q(x_k)\phi_j(x_k)]c_j=f(x_k). \qquad (3.5.3)$$
$$k=1..M$$

Example 3.5.3 Use the basis function $\phi_j=\sin j\pi x$, j=1,2,3, and the collocation points $x_k=0.25,0.5,0.75$ to compute a solution for the differential equation

$$y'' - 2(1+x)^{-2}\, y = (2x-4)/(1+x)^2, \quad y(0)=y(1)=0.$$

Solution. The trial solution for M=3 is given by

$$Y_3(x)=c_1\sin\pi x + c_2\sin 2\pi x + c_3\sin 3\pi x,$$

where the coefficients c_1, c_2 and c_3 satisfy the linear equations

$$\mathbf{Ac=f},$$

with the components of A given by $a_{kj}=(-(\pi j)^2-2/(1+x_k)^2)\sin j\pi x_k$ and those of **f** by $f_k=(2x_k-4)/(1+x_k)^2$, k=1,2,3. Solving these equations gives $c_1=0.16146304$, $c_2=0.01669060$, and $c_3=0.00450052$. This trial solution is compared with the particular solution of the differential equation satisfying the homogeneous boundary conditions in Figure 3.5.1.

Figure 3.5.1 Trial solutions Y_3 and Y_{10} for Example 3.5.1

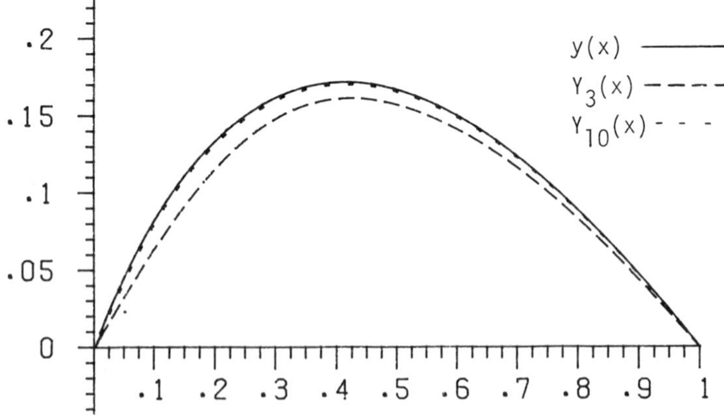

In general the matrix associated with the collocation method will be full whereas with the finite difference method it was only tridiagonal. It is possible to obtain basis functions which give a tridiagonal matrix; for example, the B-splines given by

$$B_j(x) = \begin{cases} (x-x_{j-2})^3/4h^3, & x_{j-2} \leq x \leq x_{j-1}, \\ 1/4 + 3/4h(x-x_{j-1}) + 3/4h^2(x-x_{j-1})^2 & \\ \qquad - 3/4h^3(x-x_{j-1})^3, & x_{j-1} \leq x \leq x_j, \\ 1/4 + 3/4h(x_{j+1}-x) + 3/4h^2(x_{j+1}-x)^2 & \\ \qquad -3/4h^3(x_{j+1}-x)^3, & x_j \leq x \leq x_{j+1}, \\ (x_{j+2}-x)^3/4h^3, & x_{j+1} \leq x \leq x_{j+2}, \\ 0, & \text{otherwise}, \end{cases} \tag{3.5.4}$$

where we have assumed that the points x_j are equally spaced throughout the interval (0,1). Each of these functions takes the form shown in Figure 3.5.2. In each interval (x_k, x_{k+1}), $k=j-2,..,j+1$, the function B_j is a simple cubic and the function and its first two derivatives are continuous at the points x_k, $k=j-2,..,j+2$. Outside the interval (x_{j-2}, x_{j+2}) the function is identically zero. In order to apply the collocation method we need to evaluate the B-splines, and their first two derivatives, at the points x_k, $k=1,...,M$, but, this turns out to be particularly easy as shown in Table 3.5.1.

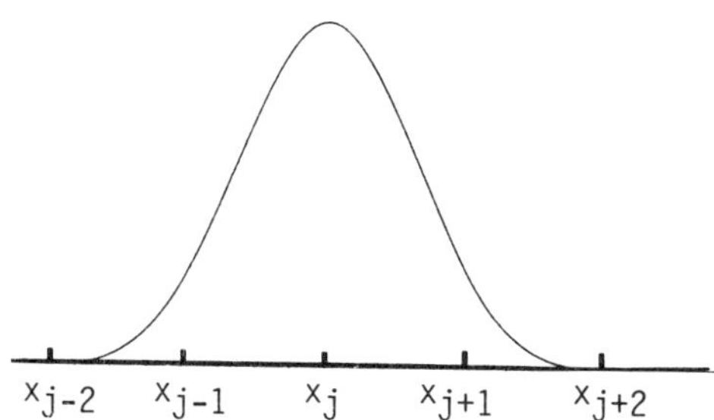

Figure 3.5.2 **The cubic B-spline $B_j(x)$.**

Table 3.5.1.	**Values of B_j, B_j' and B_j'' at x_j, $x_{j\pm1}$.**		
	x_{j-1}	x_j	x_{j+1}
$B_j(x)$	$^1/_4$	1	$^1/_4$
$B_j'(x)$	$^3/_{4h}$	0	$-^3/_{4h}$
$B_j''(x)$	$^3/_{2h^2}$	$-^3/_{h^2}$	$^3/_{2h^2}$

Substituting these expressions into (3.5.3) we obtain the approximate solution

$$y_M = \sum_{j=1..M} c_j B_j(x),$$

where

$$\mathbf{Ac = f},$$

with $c=(c_1,\ldots,c_M)^T$, $f=(f(x_1),\ldots,f(x_M))^T$. Only the three main diagonals of A are non-zero and are given by:-

$$a_{11}=-^{27}/_{2h^2}\ p(x_1)+^3/_{4h}\ p'(x_1)+^{15}/_4\ q(x_1),$$
$$a_{21}=^6/_{h^2}\ p(x_2)-^3/_h\ p'(x_2)+\ q(x_2),$$
$$a_{jj}=-^3/_{h^2}\ p(x_j)+\ q(x_j), \qquad j=2,\ldots,M-1,$$
$$a_{j,j\pm1}=^3/_{2h^2}\ p(x_j)\pm^3/_{4h}\ p'(x_j)+\ q(x_j)/4, \quad j=2,..,M-1,$$
$$a_{MM}=-^{27}/_{2h^2}\ p(x_M)-^3/_{4h}\ p'(x_M)+^{15}/_4\ q(x_M),$$
$$a_{M-1,M}=^6/_{h^2}\ p(x_{M-1})+^3/_h\ p'(x_{M-1})+\ q(x_{M-1}).$$

Exercise 3.5.1 Find an approximate solution of the differential equation $(y')'-2(1+x)^{-2}\ y=^{(2x-4)}/_{(1+x)^2}$, subject to the boundary conditions $y(0)=0$, $y(1)=0$, using cubic B-splines given by (3.5.4) and 5 equally spaced collocation points.

The choice of the collocation points, x_k, $k=1,..,M$, is as important as the choice of basis functions. We have suggested that they be evenly spaced between 0 and 1 but this is by no means essential. Other possibilities include the Gaussian quadrature points (see Davis and Rabinowitz (1975)) or the roots of the Chebyshev polynomial of degree M-1. (see Fox and Parker (1972).)

3.5.2. The Galerkin Method

Let Ω_M be the vector space generated by all linear combinations of the basis functions $\phi_j(x)$, j=1,..,M, i.e. $Y_M(x)$ is a member of Ω_M if and only if $Y_M = \sum_{j=1..M} c_j\phi_j$. For any trial solution $Y_M \in \Omega_M$ the term

$$(pY_M')' + qY_M - f = r \qquad (3.5.5)$$

is called the residual. In general r will not be a member of Ω_M; however, there is a projection mapping which applied to r gives its component in this space. Figure 3.5.3 illustrates this point for the case M=2.

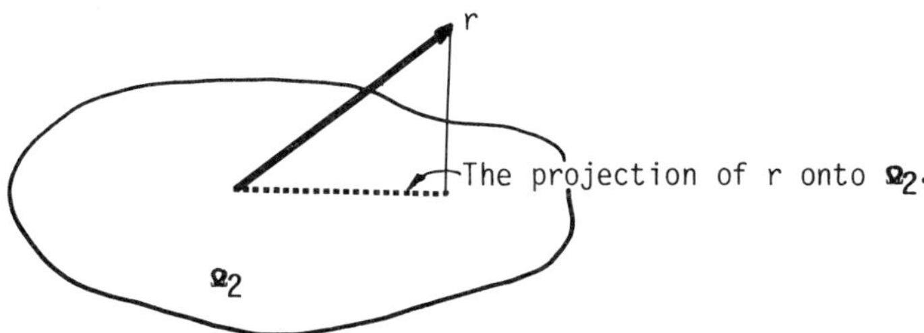

Figure 3.5.3 The projection of the residual onto Ω_2.

The projection of r onto Ω_2 is dependent upon the coefficients c_j, but it will be zero if they are selected so that r is orthogonal to each of the basis elements. Recall that two finite dimensional vectors $f=(f_1,f_2,..,f_M)^T$ and $g=(g_1,g_2,..,g_M)^T$ are said to be orthogonal if their inner product vanishes, i.e.

$$(\mathbf{f},\mathbf{g}) = \mathbf{f}^T\mathbf{g} = \sum_{j=1..M} f_j g_j = 0.$$

The corresponding definition for real valued functions is that f(x) and g(x) are orthogonal if

$$\int_0^1 f(x)g(x)dx=0.$$

This is the basis of the **Galerkin Method**; we select the coefficients c_j so that the residual is orthogonal to each of the basis functions ϕ_k, i.e.

$$\int_0^1[(pY_M')'+qY_M-f]\phi_k\, dx = 0,\ k=1,2,..,M. \quad (3.5.6)$$

Substituting the expression (3.5.2) for Y_M, interchanging the sum and the integral, gives a system of M linear equations for the coefficients c_j:

$$\mathbf{Ac} = f,$$

where $f_k=\int_0^1 f(x)\phi_k(x)dx$, k=1,..,M and

$$a_{kj} = \int_0^1[(p(x)\phi_j')'+q(x)\phi_j(x)]\phi_k dx. \quad (3.5.7)$$

We can simplify the expression (3.5.7) by integrating by parts and using the boundary conditions to give

$$a_{kj}= -\int_0^1 p(x)\phi_k'\phi_j'dx + \int_0^1 q(x)\phi_k\phi_j dx.$$

Notice that the matrix A is symmetric, but only in very simple cases will it be possible to express its entries in closed form and it may be necessary to evaluate them by numerical integration. For a general choice of basis function A will be a full matrix; however, it can be reduced to a tridiagonal form by using suitable spline functions.

Exercise 3.5.2 The linear basis functions

$$\phi_j(x) = \begin{cases} (x - x_{j-1})/h, & x_{j-1} \leqslant x \leqslant x_j, \\ (x_j - x)/h, & x_j \leqslant x \leqslant x_{j+1}, \\ 0, & \text{otherwise}, \end{cases}$$

are called the hat-functions or linear B-splines. Show that for this choice the matrix given by equation (3.5.7) reduces to a tridiagonal form.

Exercise 3.5.3 If the linear B-splines are applied to the simple problem $y''=f(x)$, $y(0)=y(1)=0$, show that the matrix A, given by (3.5.7), reduces to

$$A = {}^1/_h \begin{bmatrix} -2 & 1 & & & \\ 1 & -2 & 1 & & \\ & \ddots & \ddots & \ddots & \\ & & 1 & -2 & 1 \\ & & & 1 & -2 \end{bmatrix}$$

and that except for a factor of ${}^1/_h$ the equations for determining the coefficients c_j, $j=1,..,M$, are the same as those for determining the solution given by the finite difference method (3.4.4).

Exercise 3.5.3 demonstrates the similarity between the Galerkin and finite difference methods for the simple problem $y''=f(x)$, $y(0)=y(1)=0$. We have already seen that the finite difference method is of order 2 provided sufficient derivatives of the solution exist; precisely the same conditions are required to show that the Galerkin method using linear B-splines is of the same order. If cubic splines are used then it can be shown that the resulting approximation is order 4. Clearly, by using higher order spline functions we could improve the order of the Galerkin method, but, to do so we would increase the bandwidth of A.

3.5.3 The Rayleigh-Ritz Method

The Rayleigh-Ritz method replaces the problem of solving the self adjoint equation

$$(p(x)y')' + q(x)y = f(x), \qquad (3.5.8)$$

by the problem of minimizing a related functional. The stationary points of the functional

$$I(y) = \int_0^1 F(x,y,y')\,dx,$$

for a given F, can be determined by solving the differential equation

$$F_y - {}^d/_{dx}\, F_{y'} = 0. \qquad (3.5.9)$$

For a proof of this Variational Principle see, for example, Elsgolc (1963). Now consider the functional

$$I(y) = \int_0^1 [p(x)(y'(x))^2 - q(x)(y(x))^2 + 2yf]dx, \quad (3.5.10)$$

which we minimize amongst a set of suitably differentiable functions. From (3.5.9) the solution of this problem is given by (3.5.8). Rather than minimising (3.5.10) over all differentiable functions we only consider linear combinations of selected basis functions ϕ_j, j=1,..,M, i.e. functions which are of the form $y_M = \sum c_j\phi_j(x)$. Substituting y_M into (3.5.8) gives

$$I(c_1,...,c_M) = \int_0^1 [p(x)(\sum_{j=1..M} c_j\phi_j')^2 - q(x)(\sum_{j=1..M} c_j\phi_j')^2 - 2f(x)\sum_{j=1..M} c_j\phi_j]dx,$$

which we must minimize by selecting values for the coefficients c_j, j=1,..,M. This is an example of a finite dimensional minimization problem. For example, differentiating I with respect to each of the parameters c_j, j=1,...,M, and equating the results to zero produces a system of linear equations for the unknown coefficients. It is possible to show that if the same basis functions are selected for the Galerkin and Rayleigh-Ritz methods then the same solution is obtained. (See Strang and Fix (1973)).

Exercise 3.5.4 Use the basis functions $(1-x)x^j$, $j=1,2,3,4$, to determine an approximate solution of the differential equation

$$(y')' - 2(1+x)^{-2}\, y = (2x-4)/(1+x)^2,$$

subject to $y(0)=y(1)=0$, by (i) the Galerkin method and (ii) the Rayleigh-Ritz method. Compare the computed result with the results of Example 3.5.2. Repeat this exercise using the Galerkin and Rayleigh-Ritz methods with appropriate cubic B-spline functions.

3.5.4 Non-linear Problems

The extension of the finite element method to non-linear problems is simple; the solution of the resulting equations is, however, far from easy. Consider the differential equation

$$y''=f(x,y,y'),\quad y(0)=y(1)=0,$$

and let $Y_M = \sum_{j=1..M} c_j\phi_j(x)$. If Y_M were the solution then

$$\sum_{j=1..M} c_j\phi_j''(x) = f(x, \sum_{j=1..M} c_j\phi_j, \sum_{j=1..M} c_j\phi_j')$$

would be satisfied identically at all points in the interval (0,1). The collocation method then proceeds to determine the optimum choice for the coefficients by selecting a set of points, x_k, $k=1,..,M$ at which this expression is satisfied. This results in a system of M non-linear equations which must be solved for c_j, $j=1,..,M$. For the Galerkin method the coefficients are found by making the residual orthogonal to each of the basis functions as before, i.e.

$$\int_0^1 [Y_M'' - f(x, Y_M, Y_M')]\ \phi_k\ dx = 0.$$

Integrating the first term by parts and using the boundary conditions gives the system of non-linear equations

$$-\sum_{j=1..M} \left(\int_0^1 \phi_k' \phi_j'\ dx\right) c_j = \int_0^1 f\Big(x, \sum_{j=1..M} c_j\phi_j, \sum_{j=1..M} c_j\phi_j'\Big)\phi_k\ dx$$

for k=1,...,M. The correct choice of suitable spline functions can reduce the size of this problem but its solution in general is beyond the scope of this text. Further details are given by Strang and Fix (1973).

CHAPTER 4
Parabolic Partial Differential Equations

4.1 Introduction

Parabolic partial differential equations are exemplified by the Heat or Diffusion Equation given by $\partial u/\partial t = \sigma\, \partial^2 u/\partial x^2$. A simple introduction to the characterization of linear partial differential equations is given in Appendix 2.

We begin by giving a simple derivation of the Heat Equation. Let $u(x,y,z,t)$ be the temperature at time t at a point (x,y,z) within a homogeneous region bounded by a closed surface S. The total heat energy per unit volume is given by cu, where c is the specific heat of the region's material. If K is the thermal conductivity of this material, assumed to be constant, then heat energy flows in the direction $-\nabla u$ with magnitude $K|\nabla u|$. Let D_0 be any region, bounded by a closed surface C_0 entirely within S, then by the law of conservation of energy, the rate of change in energy within D_0 is

$$\frac{d}{dt}\iiint_{D_0} cu\, dxdydz = \iint_{C_0} K\nabla u\cdot\mathbf{n}dS,$$

where $\mathbf{n}$ is the outward unit normal to C_0 and dS is an element of area on C_0. Applying the Divergence Theorem (Gauss's Theorem) to the right-hand side of this expression and re-arranging the result gives

$$\iiint_{D_0} [\, cu_t - \nabla\cdot K\nabla u]\, dxdydz = 0$$

for any region D_0. Contracting the region D_0 to a point we conclude that the integrand must vanish, in which case, if we put $\sigma = {}^{K}/_{c}$, then

$$u_t - \sigma\nabla^2 u = 0, \qquad (4.1.1)$$

which is known as the **Heat Equation.** This equation is also called the Diffusion Equation and occurs frequently in many branches of applied mathematics and physics.

In a physical situation the solution of an equation such as (4.1.1) will be supplemented by additional conditions, for example, the initial distribution of temperature plus the temperature on the boundary of the region in which the solution is required. This is an initial boundary value or marching problem as defined in Appendix 2. Alternatively, we might require a physical solution in a region which is partly insulated, in which case the normal derivative, ${}^{\partial u}/_{\partial \mathbf{n}}$, will be specified on the corresponding part of the boundary.

If the problem is independent of y and z then equation (4.1.1) becomes

$$u_t - \sigma u_{xx} = 0, \quad \sigma > 0, \qquad (4.1.2)$$

which is the canonical form for parabolic partial diferential equations. (See Appendix 2.) In this chapter we begin by examining the solution of this equation in the region $\Omega = \{\, 0 \leqslant x \leqslant 1,\ t \geqslant 0 \}$, subject to the initial and boundary conditions:-

$$u(x,0) = \phi(x),\ 0 \leqslant x \leqslant 1, \quad \text{(Initial Condition)}, \qquad (4.1.3)$$
$$u(0,t) = u(1,t) = 0,\ t \geqslant 0, \ \text{(Boundary Conditions)}. \qquad (4.1.4)$$

Whether this problem is properly posed, i.e. whether or not it has a unique solution depending continuously on the initial and other boundary conditions, will not be considered; we shall take this as given. By using the method of separation of variables it is easy to show that any function of the form $u_n(x,t)=\sin(n\pi x)\exp(-n^2\pi^2\sigma t)$, where n is an integer, satisfies both the differential equation and the boundary conditions (4.1.4). Since the equation (4.1.2) is linear we may superpose any finite number of solutions, in which case,

$$u_N(x,t) = \sum_{j=1\ldots N} a_j u_j, \qquad (4.1.5)$$

is also a solution. Extending this solution to an infinite series requires consideration of its convergence and we will consider this point later. When t=0 the solution must reduce to $\phi(x)$, $0\leqslant x\leqslant 1$, and so

$$\phi(x) = \sum_{j=1\ldots\infty} a_j \sin j\pi x. \qquad (4.1.6)$$

We can obtain the coefficients a_j, j=1,2,..., by multiplying both sides by $\sin j\pi x$ and integrating between 0 and 1 to give

$$a_j = 2\int_0^1 \phi\,(x)\,\sin j\pi x\, dx. \qquad (4.1.7)$$

If $\phi(x)$ is sufficiently smooth then the series (4.1.6) converges. The solution of the differential equation (4.1.2) subject to the initial conditions (4.1.3) and boundary conditions (4.1.4) is then given by

$$u(x,t) = \sum_{j=1\ldots\infty} a_j \sin j\pi x \exp(-j^2\pi^2\sigma t), \qquad (4.1.8)$$

where a_j is given by (4.1.7). This series converges for $t>0$ by comparing it term by term with the convergent series (4.1.6).

Although the solution of problem (4.1.2)-(4.1.4) is given by the series (4.1.8) this expression is not very convenient. The series may converge very slowly and many hundreds of terms may be required before a solution of sufficient accuracy is found. Furthermore, this technique does not generalise to other boundary conditions nor to equations where $\sigma=\sigma(x,t)$ or $\sigma=\sigma(x,t,u)$. However, we shall use equation (4.1.2) to develop numerical techniques which will extend to more difficult problems.

4.2 Simple Numerical Methods for Solving the Heat Equation $u_t=\sigma u_{xx}$

We seek a solution of the equation $u_t=\sigma u_{xx}$, throughout the region $\Omega = (0,1)\times(0,T)$, which satisfies the boundary and initial conditions (4.1.3) and (4.1.4). The region Ω is covered by a mesh of lines which are spaced h apart in the x direction and k apart in the t direction; only where these lines meet will an attempt be made to approximate the solution.

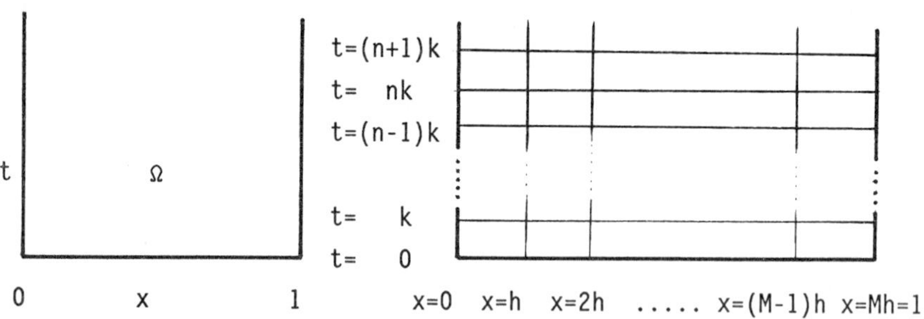

Figure 4.2.1 The region Ω and its covering mesh $\Omega_{h,k}$.

Let U_m^n be a computed difference approximation for u(mh,nk) at each point of the mesh $\Omega_{h,k}=\{(x,t) : x=mh, 0<m<M, Mh=1, t=nk, n>0\}$. This difference solution is said to converge to the solution of the differential equation at a point (x,t) if $|u(x,t)-U_m^n| \to 0$ as $h,k\to 0$, $n,m\to\infty$ with mh=x and nk=t.

The initial condition (4.1.3) provides values for $U_m^0=\phi(mh)$, m=1,2,..,M-1, and the homogeneous boundary conditions (4.1.4) give $U_0^n=U_M^n=0$, for n=1,2,.... Next we replace the differential equation $u_t=\sigma u_{xx}$ by a suitable difference approximation and use the result to advance the solution through the mesh $\Omega_{h,k}$. For example, if we replace $\partial^2u/\partial x^2$ by $[u(x+h,t)-2u(x,t)+u(x-h,t)]/_{h^2}$ and $\partial u/\partial t$ by $[u(x,t+k)-u(x,t)]/_k$ then

$$\frac{u(x,t+k)-u(x,t)}{k} = \sigma\,\frac{u(x+h,t)-2u(x,t)+u(x-h,t)}{h^2} + E(x,t), \quad (4.2.1)$$

where $E(x,t)$ is the local discretization error caused by such an approximation. Expanding each term in a Taylor series about the point (x,t) and assuming that sufficient derivatives of u exist then the local discretization error is given by

$$E(x,t)=(u_t-\sigma u_{xx})+{}^{k}/_{2}\, u_{tt}-{}^{h^2\sigma}/_{12}\, u_{xxxx} + \text{Higher order terms.}$$

Clearly, the first term vanishes since we are trying to solve $u_t=\sigma u_{xx}$ and so the discretization error is $O(k+h^2)$. If the mesh sizes, h and k, are sufficiently small we may neglect the local discretization error, replace $u(mh,nk)$ by $U_m^{\ n}$ and obtain the difference method

$$U_m^{\ n+1} = U_m^{\ n} + {}^{k\sigma}/_{h^2}[U_{m+1}^{\ n}-2U_m^{\ n}+U_{m-1}^{\ n}], \quad (4.2.2)$$
$$m=1,2,\ldots,M-1,\ n>0.$$

This is called an explicit method since only one point at the $n+1^{th}$ level of the mesh $\Omega_{h,k}$ is determined at any one time. As the local discretization error tends to zero as the mesh is refined, i.e. as h and k tend to zero, the method (4.2.2) is said to be consistent with the differential equation $u_t=\sigma u_{xx}$. Using the central difference operator $\delta_x u(x,t)=[u(x+{}^{h}/_{2},t)-u(x-{}^{h}/_{2},t)]$, equation (4.2.2) may also be written in the compact form

$$U_m^{\ n+1} = U_m^{\ n} + k\ [{}^{\sigma}/_{h^2}\ \delta_x^2 U_m^{\ n}], \qquad (4.2.3)$$
$$m=1,2,\ldots,M-1,\ n>0.$$

We can compare this approach with the application of Euler's method to a single ordinary differential equation as follows. If A denotes the operator ${}^{\partial^2}/_{\partial x^2}$, the equation $u_t=\sigma u_{xx}$ may be written as

$${}^{du}/_{dt} = \sigma Au, \qquad (4.2.4)$$

for which Euler's method produces $U^{n+1}=U^n+k[\sigma AU^n]$. We have seen that consistency is insufficient to guarantee that Euler's method will converge as the step size $k\to 0$; consistency merely means that the equation (4.2.1) tends to the differential equation $u_t=\sigma u_{xx}$ as h and k approach zero. In addition, we need to consider the stability of the difference method, i.e. if a small error should occur will the consequences of such a perturbation swamp the solution we are trying to obtain? We shall consider this problem after a simple example which demonstrates how the difference method (4.2.2) is applied.

Example 4.2.1 Compute a difference solution for $u_t=u_{xx}$, subject to the boundary conditions $u(0,t)=u(1,t)=0$ and the initial condition $u(x,0)=\sin\pi x$ using h=0.1 and k=0.005. Repeat this problem for h=0.05 and k=0.0025.

Solution. The region $\Omega=(0,1)\times(0,T)$ is covered with a mesh which has spacing h=0.1, (hence M=10 since Mh=10×0.1=1), and k=0.005. The initial conditions give $U_m^0=u(mh,0)=\sin\pi hm$, m=0,1,...,10. We now apply the difference equation (4.2.2) with $r=k\sigma/h^2=0.005/0.1^2$, i.e. r=0.5. In particular when n=0 and m=1 equation (4.2.2) becomes

$$U_1^1 = U_1^0 + 0.5[U_2^0 - 2U_1^0 + U_0^0].$$

But as shown in Figure 4.2.2 the values U_0^0, U_1^0 and U_2^0 are known initial values and so we can calculate U_1^1. In the same way using n=0, m=2 we can find U_2^1, then U_3^1 and so on until $\{U_m^1\}$, m=1,2,..,8,9 are determined. The boundary conditions then provide values for U_0^1 and U_{10}^1. Similarly, we can find $\{U_m^2\}$, m=0,1,2,...,9,10 and proceed to compute a solution over the mesh $\Omega_{h,k}$.

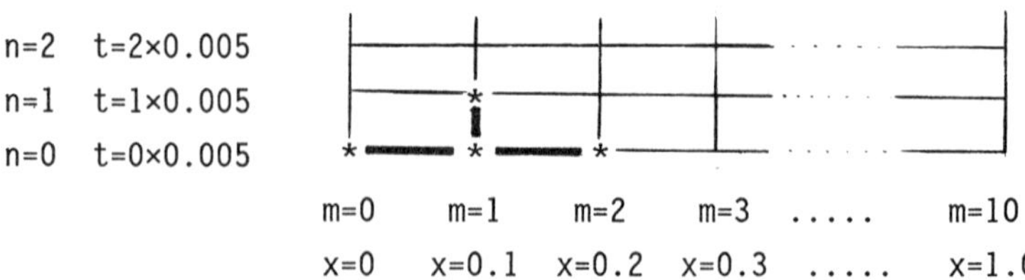

Figure 4.2.2. Diagrammatic representation of the scheme 4.2.2

Table 4.2.1 The Computed Solution Determined for Example 4.2.1

	t=0.02			t=0.04		
x	(a)	(b)	(c)	(a)	(b)	(c)
0.00	0.000000	0.000000	0.000000	0.000000	0.000000	0.000000
0.05	0.128412		0.128149	0.105409		0.104999
0.10	0.253662	0.252818	0.253140	0.208224	0.206839	0.207328
0.15	0.372667		0.371899	0.305910		0.304698
0.20	0.482495	0.480888	0.481501	0.396065	0.393432	0.394391
0.25	0.580442		0.579247	0.476467		0.474546
0.30	0.664097	0.661886	0.662730	0.545136	0.541512	0.542865
0.35	0.731399		0.729894	0.600383		0.597918
0.40	0.780693	0.778093	0.779086	0.640846	0.636586	0.638370
0.45	0.810762		0.809094	0.665530		0.662610
0.50	0.820869	0.818136	0.819179	0.673825	0.669346	0.671428
0.55	0.810762		0.809094	0.665530		0.662610
0.60	0.780693	0.778093	0.779086	0.640846	0.636586	0.638370
0.65	0.731399		0.729894	0.600383		0.597918
0.70	0.664097	0.661886	0.662730	0.545136	0.541512	0.542865
0.75	0.580442		0.579247	0.476467		0.474546
0.80	0.482495	0.480888	0.481501	0.396065	0.393432	0.394804
0.85	0.372667		0.371899	0.305910		0.304552
0.90	0.253662	0.252818	0.253140	0.208224	0.206839	0.207309
0.95	0.128412		0.128148	0.105409		0.105053
1.00	0.000000	0.000000	0.000000	0.000000	0.000000	0.000000

	t=0.06			t=0.08		
x	(a)	(b)	(c)	(a)	(b)	(c)
0.00	0.000000	0.000000	0.000000	0.000000	0.000000	0.000000
0.05	0.086527		0.166357	0.071028		200.2461
0.10	0.170924	0.169223	0.026516	0.140306	0.138447	-264.7085
0.15	0.251112		0.420760	0.206130		53.85112
0.20	0.325117	0.321880	0.179793	0.266878	0.263342	578.1252
0.25	0.391116		0.421332	0.321055		-1765.833
0.30	0.447485	0.443030	0.640548	0.367327	0.362459	3613.145
0.35	0.492836		-0.088526	0.404553		-6143.550
0.40	0.526051	0.520814	1.658603	0.431818	0.426096	9266.890
0.45	0.546312		-1.307606	0.448451		-12737.54
0.50	0.553122	0.547616	3.197644	0.454041	0.448024	16172.77
0.55	0.546312		-2.853685	0.448451		-19092.50
0.60	0.526051	0.520814	4.459284	0.431818	0.426096	21031.20
0.65	0.492836		-3.639206	0.404553		-21640.29
0.70	0.447485	0.443030	4.356571	0.367327	0.362459	20796.73
0.75	0.391116		-2.941717	0.321055		-18623.45
0.80	0.325117	0.321880	2.850369	0.266878	0.263342	15454.47
0.85	0.251112		-1.438382	0.206130		-11709.70
0.90	0.170924	0.169223	1.135697	0.140306	0.138447	7766.410
0.95	0.086527		-0.334337	0.071028		-3847.470
1.00	0.000000	0.000000	0.000000	0.000000	0.000000	0.000000

(a) The value of the exact solution $\exp(-\pi^2 t)\sin\pi x$.

(b) Difference solution with h=0.1, k=0.005, r=0.5.

(c) Difference solution with h=0.05, k=0.0025, r=1.

Figure 4.2.3 **The Computed Solution of $u_t=u_{xx}$ corresponding to Example 4.2.1.**

Exact solution $\exp(-\pi^2 t)\sin\pi x$, + r=0.5, • r=1.0

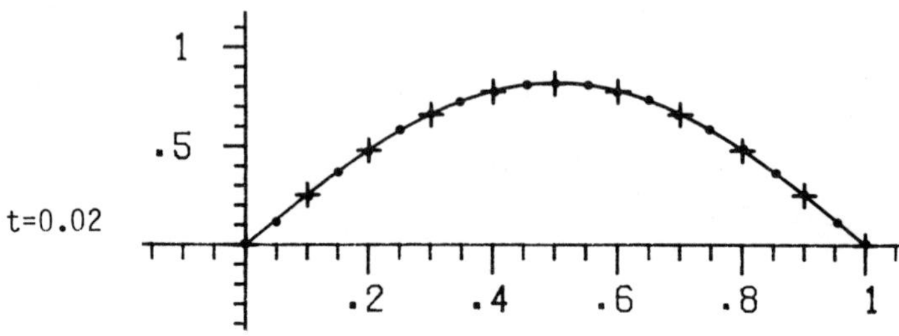

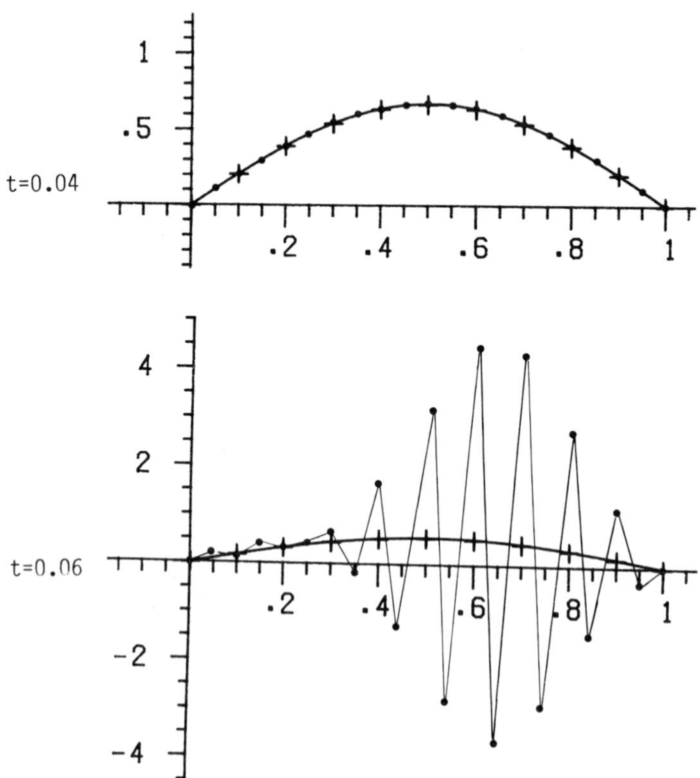

The computed solution for Example 4.2.1 is given in Table 4.2.1 and Figure 4.2.3 at intervals of t=0.02. For h=0.1 and k=0.005 the computed solution agrees quite well, both qualitatively and quantitatively. Halving both h and k produces a difference solution which grows rapidly and oscillates, and is clearly unstable.

The last example demonstrates that stability is necessary if the difference solution is to resemble the solution of the differential equation. Let u(mh,nk) be the solution of the differential equation at x=mh, t=nk for which we compute a value U_m^n using the difference equation (4.2.2). However, U_m^n will not be the exact solution of the difference equation because it will be contaminated by various unavoidable errors. (For example, in representing the coefficients.) Therefore, if V_m^n is the correct solution of the difference equation then

$$|u(mh,nk)-U_m^n| \leq |u(mh,nk)-V_m^n| + |V_m^n-U_m^n|. \quad (4.2.5)$$

The first term on the right-hand side can be controlled by selecting step sizes h and k so that the local discretization error is sufficiently small. The second term is a measure of the stability of the difference scheme. Accordingly, we require a condition which will guarantee the stability of the difference solution. Recall that this difference solution is obtained by applying the difference method (4.2.2) for m=1,2,..,M-1, all of which may be collected together to give

$$\begin{bmatrix} U_1^{n+1} \\ U_2^{n+1} \\ \cdot \\ \cdot \\ \cdot \\ U_{M-1}^{n+1} \end{bmatrix} = \begin{bmatrix} 1-2r & r & 0 & & & 0 \\ r & 1-2r & r & 0 & & 0 \\ 0 & \cdot & \cdot & \cdot & & \cdot \\ \cdot & & \cdot & \cdot & \cdot & \cdot \\ \cdot & & & r & 1-2r & r \\ 0 & & & 0 & r & 1-2r \end{bmatrix} \begin{bmatrix} U_1^n \\ U_2^n \\ \cdot \\ \cdot \\ \cdot \\ U_{M-1}^n \end{bmatrix} + \begin{bmatrix} rU_0^n \\ 0 \\ \cdot \\ \cdot \\ 0 \\ rU_M^n \end{bmatrix} \quad (4.2.6)$$

Using the boundary conditions (4.1.4) produces $U_0^n=U_M^n=0$ and so

$$U^{n+1} = \mathbf{C}U^n, \qquad (4.2.7)$$

where $\mathbf{U}^n=[U_1{}^n,U_2{}^n,..,U_{M-1}{}^n]^T$ and $\mathbf{C}$ is the tridiagonal matrix as given in equation (4.2.6). Equation (4.2.7) is a simple matrix recurrence relation for which we consider stability subject to an initial perturbation. Suppose that there is an error ε^0 in $\mathbf{U}^0$, then the error in $\mathbf{U}^1$ will be $[\mathbf{C}(\mathbf{U}^0+\varepsilon^0)]-[\mathbf{C}\mathbf{U}^0]=\mathbf{C}\varepsilon^0$, which becomes $\mathbf{C}^2\varepsilon^0$ in $\mathbf{U}^2$ and $\mathbf{C}^n\varepsilon^0$ in $\mathbf{U}^n$. If $\|\mathbf{C}\|\leq 1$ then this error will remain bounded as $n\to\infty$. From Appendix 1 we see that the eigenvalue of $\mathbf{C}$ with largest modulus is given by $1-2r+2\sqrt{r^2}\cos(\pi/M)=1-2r(1-\cos\pi/M)$ and so $\|\mathbf{C}\|_2\leq 1$ provided $r=k\sigma/h^2\leq 1/2$; accordingly, the method (4.2.2) is said to be conditionally stable. For Example 4.2.1, the choice h=0.1, k=0.005 gives r=0.5 which which satisfies the above stability condition. Halving both h and k gives a value for r=1 and the resulting difference solution is unstable, growing rapidly with a characteristic oscillatory nature. In order to retain stability when h is halved it is necessary to reduce k by a factor of 4. We now show how the condition $r\leq 1/2$ may be used to help prove that the difference solution converges as h and k → 0. We write equation (4.2.1) as

$$u(x,t+k)=u(x,t) + r(u(x+h,t)-2u(x,t)+u(x-h,t))+kE(x,t), \qquad r=k\sigma/h^2,$$

and the difference equation (4.2.2) as

$$U_m{}^{n+1} = U_m{}^n + r(U_{m+1}{}^n-2U_m{}^n+U_{m-1}{}^n), \qquad m=1,2,..,M-1.$$

At each point of the mesh in $\Omega_{h,k}$ we set $e_m{}^n=u(mh,nk)-U_m{}^n$ then

$$e_m{}^{n+1} = e_m{}^n + r(e_{m+1}{}^n-2e_m{}^n+e_{m-1}{}^n) + kE(x,t), \; m=1,2,..,M-1.$$

If $E_n=\max|e_m{}^n|$, m=1,2,..,M-1, and $E=\sup|E(x,t)|$, (x,t) in Ω, then

$$\begin{aligned}|e_m{}^{n+1}| &\leq r|e_{m+1}{}^n|+|1-2r|\;|e_m{}^n|+r|e_{m-1}{}^n| + kE\\ &\leq rE_n + |1-2r|E_n + rE_n + kE,\end{aligned}$$

which, if $r \leq {}^1/_2$, gives

$$|e_m^{n+1}| \leq E_n + kE$$

and so

$$E_{n+1} \leq E_n + kE.$$

Applying this result repeatedly gives

$$E_n \leq E_{n-1} + kE \leq E_{n-2} + 2kE \leq \ldots . \leq E_0 + nkE = E_0 + tE.$$

Therefore, $E_n \to 0$ provided the discretization error bound E and E_0 tend to zero as the mesh is refined. The term E_0 is a measure of the accuracy with which we can evaluate the initial condition and may be made as small as we please. Finally, as $E_n = \max|u(mh,nk) - U_m^n|$, $m=1,\ldots,M-1$, the difference solution converges to the solution of equation (4.1.2) as the mesh is refined. Fortunately, there is no need to prove a similar result for each difference method which only involves two levels of $\Omega_{h,k}$ because a general theorem is available.

Theorem 4.2.1 (The Lax Equivalence Theorem.) Given a properly posed initial value problem and a difference method which is consistent with it, stability is a necessary and sufficient condition for convergence.

By properly posed problem we shall mean that the solution of the differential equation depends continuously and uniquely on the initial data. A comprehensive proof is given by Richtmyer and Morton (1967) in terms of the general setting of Banach spaces. Notice that the Lax Equivalence Theorem not only says that if a given difference method is consistent and stable then it will converge, but also that any difference method which is unstable cannot converge.

Stability as discussed previously relies heavily upon a matrix analysis which does not generalise in a simple way. We now consider an alternative technique. For stability it is necessary that the solution of the difference method should remain bounded, for if it does not then it cannot be a reasonable approximation for the solution which is known to be bounded. Therefore, we shall examine the solution of the difference equation (4.2.2) directly by assuming it has the form of an exponential Fourier series, i.e. $\sum A_n \rho^n \exp(im\psi)$, $i=\sqrt{-1}$. This expression may be compared with the solution of the differential equation (4.1.2), which is given by equation (4.1.8), i.e. $u(x,t)=\sum a_j \sin j\pi x \exp(-j^2\pi^2\sigma t)$, taking $\rho=\exp(-j^2\pi^2\sigma k)$, $\psi=j\pi h$, expressing $\sin j\pi x$ in terms of exponential functions and re-ordering the index of the summation. (We have assumed that the boundary conditions are homogeneous; if not, then provided simple boundary conditions $u(0,t)=\alpha$, $u(1,t)=\beta$ are given we can subtract a suitable linear function, $f(x)=(\beta-\alpha)x+\alpha$, from the solution to produce homogeneous boundary conditions. For more complicated boundary conditions we have to assume that we are sufficiently far from the boundary for such an analysis to be valid. This is the major drawback with this approach; the matrix stability analysis takes the boundary conditions into account automatically.) In fact because of the linearity of the difference equation (4.2.2) we need only consider each individual term of the Fourier series and then superpose them. Accordingly, let us look for a solution of equation (4.2.2) which is of the form $U_m^n=A\rho^n\exp(im\psi)$. This expression must satisfy the difference equation identically and so

$$A\rho^{n+1}e^{im\psi} = A\rho^n e^{im\psi} + r(A\rho e^{i[m+1]\psi}-2A\rho^n e^{im\psi}+A\rho^n\, e^{i[m-1]\psi}).$$

Clearly A is arbitrary and we can cancel a term $\rho^n e^{im\psi}$. Therefore ρ, which is called the **amplification factor**, is given by

$$\rho=1+r(e^{i\psi}-2+e^{-i\psi})=1-2r(1-\cos\psi).$$

Since $|e^{im\psi}|=1$ the difference solution $U_m^n=A\rho^n e^{im\psi}$ will remain bounded provided $|\rho|\leqslant 1$, therefore, we require

$$-1\leqslant 1-2r(1-\cos\psi)\leqslant 1, \quad \text{for all values of } \psi,$$

or $r\leqslant {}^1/_2$ as before. A justification of this type of analysis, due to von Neumann, is given by Richtmyer and Morton (1967).

The simple explicit difference method (4.2.2) has local discretization error $O(k+h^2)$ which means that unless very small steps are taken the difference solution which is produced will be very crude. We will consider now some alternative explicit difference methods which are more accurate.

Example 4.2.2 The simple explicit method (4.2.2) has a local discretization error $E={}^{k}/_{2}\, u_{tt}-{}^{h^2\sigma}/_{12}\, u_{xxxx}+O(k^2)+O(h^4)$; show that if σ is constant then $r={}^{1}/_{6}$ gives a method which has $E=O(k^2+h^4)$.

Solution Since σ is constant $u_{tt}=(u_t)_t=(\sigma u_{xx})_t=\sigma(u_t)_{xx}=\sigma^2 u_{xxxx}$ and so $E=({}^{k\sigma}/_{2}-{}^{h^2}/_{12})\sigma u_{xxxx}+O(k^2)+O(h^4)$. The first term of this expression will vanish provided $r={}^{k\sigma}/_{h^2}={}^{1}/_{6}$. Notice that this value of r gives a stable scheme but only allows a step ${}^{1}/_{3}$ of that permitted by the simple explicit scheme. This method is usually associated with Douglas.

Exercise 4.2.1 An alternative means of improving the accuracy of this type of difference method is to replace the derivatives by higher order difference approximations. For example, if we replace ${}^{\partial u}/_{\partial t}$ by the centred difference ${}^{[u(x,t+k)-u(x,t-k)]}/_{2k}$ we obtain Richardson's explicit difference method

$$U_m^{n+1} = U_m^{n-1} + {}^{2k\sigma}/_{h^2}[U_{m+1}^{n}-2U_m^{n}+U_{m-1}^{n}], \qquad (4.2.8)$$
$$m=1,2,\ldots,M-1, \quad n>0.$$

Show that the local discretization error for this scheme is $O(k^2+h^2)$ but that this method is **unconditionally unstable**. This method is explicit but involves three levels of the mesh and so a special starting procedure is required. Notice the similarity between this method and the application of equation (2.7.1) with (2.7.2) to the ordinary differential equation ${}^{dy}/_{dx}=f(x,y)$, $y(a)=\alpha$, which has a similar outcome.

Exercise 4.2.2 Show that if the term U_m^n in (4.2.8) is replaced by $[U_m^{n+1}+U_m^{n-1}]/_2$ then the corresponding method is stable. Show also that the local discretization error is $O(k^2+h^2+[k/h]^2)$. This method is usually attributed to DuFort and Frankel. Notice that the discretization error tends to zero as h and k → 0 only if k/h also tends to zero. Use the von Neumann analysis to show that this method is unconditionally stable.

The simple explicit method (4.2.3) involves 4 points, only one of which is at the n+1th time level, and is very simple to use; however, the stability condition ($k\sigma/h^2 \leq 0.5$) is very restrictive. The difference scheme (4.2.8) involves 5 points and is unstable, but it can be modified using the same 5 points to give a method which is unconditionally stable. However, for consistency we require that k tends to zero faster than h, i.e. $k \propto h^{1+\varepsilon}$, with $\varepsilon>0$. Clearly, by including more points we can obtain more accurate difference methods. If such methods include points at the n-1th time level then they are called multi-step and may suffer from induced parasitic solutions just as with ordinary differential equations; this is precisely what causes the unstable nature of equation (4.2.8). Only if all the parasitic solutions are bounded can the method be stable. Although Theorem 4.2.1 does not apply directly to multi-level schemes stability is clearly a necessary condition for convergence.

An alternative means of improving the accuracy of the computed solution is to use the extrapolation techniques used in the previous chapter. For example, if we compute a solution at the n+1th time level using a step size k, i.e.

$$V_m^{n+1} = U_m^n + r\delta^2 U_m^n, \qquad (4.2.9)$$

and then carry out two steps with a step size $k/2$, i.e.

$$W_m^{n+1/2} = U_m^n + r/2\ \delta^2 U_m^n, \qquad (4.2.10)(i)$$
$$W_m^n = W_m^{n+1/2} + r/2\ \delta^2 W_m^{n+1/2}, \qquad (4.2.10)(ii)$$

then, based upon the fact that the local discretization error is $O(k+h^2)$, we may extrapolate the values V_m^{n+1} and W_m^{n+1} to obtain

$U_m^{n+1}=2W_m^{n+1}-V_m^{n+1}$. Eliminating W_m^n and V_m^n produces

$$U_m^{n+1} = U_m^n + r\delta^2 U_m^n + {}^r/_2\delta^4 U_m^n, \qquad (4.2.11)$$

which has local error $O(k^2+h^2)$. Notice that, if σ is constant, the term ${}^r/_2\delta^4 U_m^n$ eliminates ${}^k/_2\ {}^{\partial^2 u}/_{\partial t^2} = {}^k/_2\ \sigma^2\, {}^{\partial^4 u}/_{\partial x^4}$ from the local discretization error of the simple explicit method (4.2.2). This produces a difference scheme which is similar to Douglas's method as given in Example 4.2.2. The method (4.2.11) is applied in the form (4.2.8)-(4.2.10) since each of these equations involves only 4 points whereas the method (4.2.11) involves 5 points at the nth level plus one at the $n+1^{th}$ level of the mesh. Furthermore, the use of 5 points at the nth level of the mesh causes difficulty in determining the solution near the boundaries at $x=0$ and $x=1$. In fact the composite method (4.2.11) is nothing more than the Euler-Cauchy Runge-Kutta method given by equation (2.5.3) applied to the differential equation (4.2.4). A simple application of the von Neumann analysis gives stabilty provided ${}^{k\sigma}/_{h^2}\leq 0.5$ which is no improvement on the simple explicit method.

Exercise 4.2.3 Carry out the von Neumann stability analysis for the method (4.2.11) and confirm that this method is stable provided ${}^{k\sigma}/_{h^2}\leq 0.5$.

The rate at which the difference solution converges to the solution of the differential equation depends not only upon the accuracy of the difference method but also upon the behaviour of the boundary and initial conditions. For example, let us consider the solution of $u_t=\sigma u_{xx}$, subject to homogeneous boundary conditions at $x=0$ and $x=1$, and initial condition $u(x,0)=\phi(x)$, $0\leq x\leq 1$, where ϕ and its first $p-1$ derivatives are supposed continuous but its pth derivative has only bounded variation. If the simple explicit method (4.2.2) were used with $k={}^{\sigma h^2}/2$, then the rate at which the calculated difference solution converges to the analytic solution of the problem is

$$U_m^{\ n} - u(mh,nk) = \begin{cases} O(k^{p/4}) = O(h^{p/2}), & \text{for } p\geq 3, \\ O(k|\log_e k|) = O(h^2|\log_e h|), & \text{for } p=4 \\ O(k) = O(h^2), & \text{for } p>4. \end{cases}$$

Accordingly, if ϕ is a piecewise linear function then p=2 and the above error tends to zero like $\sqrt{h}$ as $h\to 0$. If the more accurate explicit method due to Douglas, i.e. $k={}^{\sigma h^2}/_6$, is used then it can be shown that

$$U_m^{\ n} - u(mh,nk) = \begin{cases} O(k^{p/3}), & \text{for } p\leq 5, \\ O(k^2|\log_e k|), & \text{for } p=6, \\ O(k^2), & \text{for } p>6. \end{cases}$$

Notice that if the initial function is only piecewise linear then there is little advantage to be gained in using the more accurate scheme, but if ϕ is sufficiently smooth then the rate of convergence increases from $O(k)$ to $O(k^2)$.

As with ordinary differential equations,if we wish to improve the accuracy of the methods available and also reduce any restrictions due to possible instability it is necessary to resort to implicit methods.

4.3 Implicit Methods for Solving the Heat Equation $u_t=\sigma u_{xx}$

The simplest implicit method for obtaining a difference solution for the ordinary differential equation $y'=f(x,y)$, subject to $y(0)=\alpha$, is given by the Trapezium Rule

$$Y^{n+1} = Y^n + {}^k/_2\ [f(x_n,Y^n) + f(x_{n+1},Y^{n+1})],$$

where $Y^n \simeq y(nk)$. If we consider the partial differential equation (4.1.2) in the form ${}^{du}/_{dt}=\sigma Au$ then the Trapezium method gives

$$U^{n+1} = U^n + {}^{k\sigma}/_2[\, AU^{n+1} + AU^n]$$

which becomes

$$U_m^{n+1} = U_m^n + 0.5r\ [\delta_x^2 U_m^{n+1} + \delta_x^2 U_m^n],\ r={}^{k\sigma}/_{h^2}, \qquad (4.3.1)$$

or

$$U_m^{n+1} - 0.5r[U_{m+1}^{n+1}-2U_m^{n+1}+U_{m-1}^{n+1}] = U_m^n + 0.5r[U_{m+1}^n-2U_m^n+U_{m-1}^n], \qquad (4.3.2)$$
$$m=1,2,\ldots,M-1,\quad n>0.$$

This is an implicit method since it involves more than one point at the n+1st time level of the mesh, U_{m+1}^{n+1}, U_m^{n+1} and U_{m-1}^{n+1}; therefore, we cannot determine each point in isolation as with the simple explicit method (4.2.2). Instead we write out all M-1 equations given by (4.3.2) and set $U_0^{n+1}=U_M^{n+1}=0$ to satisfy the homogeneous boundary conditions (4.1.3). This gives the system of linear equations

$$\begin{bmatrix} 1+r & -r/2 & 0 & \cdot & & 0 \\ -r/2 & 1+r & -r/2 & 0 & \cdot & 0 \\ 0 & \cdot & \cdot & \cdot & & \cdot \\ \cdot & \cdot & \cdot & \cdot & & \cdot \\ \cdot & & 0 & -r/2 & 1+r & -r/2 \\ 0 & \cdot & \cdot & 0 & -r/2 & 1+r \end{bmatrix} \begin{bmatrix} U_1^{n+1} \\ U_2^{n+1} \\ \cdot \\ \cdot \\ \cdot \\ U_{M-1}^{n+1} \end{bmatrix} = \begin{bmatrix} 1-r & r/2 & 0 & \cdot & & 0 \\ r/2 & 1-r & r/2 & 0 & \cdot & 0 \\ 0 & \cdot & \cdot & \cdot & & \cdot \\ \cdot & \cdot & \cdot & \cdot & & \cdot \\ \cdot & & 0 & r/2 & 1-r & r/2 \\ 0 & \cdot & \cdot & 0 & r/2 & 1-r \end{bmatrix} \begin{bmatrix} U_1^n \\ U_2^n \\ \cdot \\ \cdot \\ \cdot \\ U_{M-1}^n \end{bmatrix}$$

which may be written as

$$C_1U^{n+1} = C_2U^n, \qquad (4.3.3)$$

in an obvious notation. To determine the difference solution at the n+1st level of the mesh, i.e. $U^{n+1}=[U_1^{n+1},U_2^{n+1},\ldots,U_{M-1}^{n+1}]^T$, it is necessary to solve this system of equations. Notice that these equations are tridiagonal and may be solved using a stable LU decomposition without the need for partial pivoting. Furthermore, the matrix C_1 is the same at each level of the mesh and so we need carry out only one decomposition; the solution is then given at each step by a forward and backward substitution. (See Section 1.4.)

Exercise 4.3.1 Compute a difference solution for $u_t=u_{xx}$, subject to the initial condition $u(x,0)=\sin\pi x$, $0\leq x\leq 1$, and boundary conditions $u(0,t)=u(1,t)=0$, with $h=0.1$ and $k=0.005$. Repeat the computation using halved step sizes and compare the calculated difference solutions with Example 4.2.1.

Equation (4.3.1) is a special case of the more general implicit method

$$U_m^{n+1} = U_m^n + r[\theta\delta_x^2 U_m^{n+1}+(1-\theta)\delta_x^2 U_m^n], \qquad (4.3.4)$$
$$m=1,2,\ldots,M-1, \quad n>0.$$

For $\theta={}^1/_2$ equation (4.3.4) reduces to equation (4.3.1) and is called the Crank-Nicolson method. The choice $\theta=1$ gives the simple explicit method (4.2.2) whilst the choice $\theta=0$ gives what is called a fully implicit method for the obvious reason. The discretization error associated with equation (4.3.4) can be determined as follows. Recall that $U_m^n\simeq u(mh,nk)$ and so (4.3.4) can be compared with

$$\frac{u(x,t+k)-u(x,t)}{k} = \sigma\Big\{\theta\Big(\frac{u(x+h,t+k)-2u(x,t+k)+u(x-h,t+k)}{h^2}\Big) + (1-\theta)\Big(\frac{u(x+h,t)-2u(x,t)+u(x-h,t)}{h^2}\Big)\Big\} + E$$

where $x=mh$, $t=nk$. If $u(x,t)$ is sufficiently differentiable and σ is constant then the local discretization error, E, is given by

$$E(x,t) = \sigma[\sigma({}^1/_2-\theta)k - {}^{h^2}/_{12}]{}^{\partial^4 u}/_{\partial x^4} + O(k^2+h^4). \qquad (4.3.5)$$

In general $E=O(k+h^2)\to 0$ as h and k tend to zero and so the difference method (4.3.4) is consistent with the differential equation (4.1.2). For the Crank-Nicolson method, given by equation (4.3.2), $E=O(k^2+h^2)$. Notice that $\theta={}^1/_2-{}^{h^2}/_{12\sigma k}$ gives a method with local discretization error $O(k^2+h^4)$.

Exercise 4.3.2 Confirm that the local discretization error associated with the difference method (4.3.4) is given by equation (4.3.5). (Hint: Expand in a Taylor series about $(x,t+{}^k/_2)$.)

In order to show that the difference solution produced by the method (4.3.4) converges to the solution of the equation $u_t=\sigma u_{xx}$, subject to (4.1.3) and (4.1.4), we need only establish that this consistent method is also stable and apply Lax's Equivalence Theorem. To do this we write the equation (4.3.3) as

$$U^{n+1} = \mathbf{C}_1^{-1}\mathbf{C}_2 U^n, \qquad (4.3.6)$$

and note that $\mathbf{C}_1=\mathbf{I} - 0.5r\theta\mathbf{C}$ and $\mathbf{C}_2=\mathbf{I} - 0.5r(1-\theta)\mathbf{C}$, where

$$\mathbf{C} = \begin{bmatrix} -2 & 1 & 0 & & & & \\ 1 & -2 & 1 & 0 & & & \\ & & \cdot & \cdot & \cdot & & \\ & & & 0 & 1 & -2 & 1 \\ & & & & 0 & 1 & -2 \end{bmatrix}. \qquad (4.3.7)$$

The vector recurrence relation (4.3.6), and hence the difference method (4.3.4), is stable if and only if $\|C_1^{-1}C_2\| \leqslant 1$. Notice that C_1 is non-singular but we emphasise that C_1^{-1} is **not explicitly determined**; the difference solution U^{n+1} is determined by solving the equations (4.3.3). Now we find the eigenvalues of $C_1^{-1}C_2$ and note that if x_i is an eigenvector of C with eigenvalue λ_i then

$$\mathbf{C}_1\mathbf{x}_i=(\mathbf{I}-0.5r\theta\mathbf{C})\mathbf{x}_i=(1-0.5r\theta\lambda_i)\mathbf{x}_i,$$

and

$$\mathbf{C}_2\mathbf{x}_i=(\mathbf{I}+0.5r(1-\theta)\mathbf{C})\mathbf{x}_i=(1+0.5r(1-\theta)\lambda_i)\mathbf{x}_i$$

and so the eigenvectors of $C_1^{-1}C_2$ are also given by $\mathbf{x}_i$ with the corresponding eigenvalues given by

$$(1+0.5r(1-\theta)\lambda_i) \;/\; (1 - 0.5r\theta\lambda_i),$$

where $\lambda_i=-2+2\cos{}^{i\pi}/_{M}$, $i=1,2,\ldots,M$. Since both C_1 and C_2 are symmetric matrices

$$\|C_1^{-1}C_2\|_2 = \underset{1\leqslant i\leqslant M-1}{\text{maximum}} \left|\frac{(1 - 0.5r(1-\theta)\sin^2\phi_i)}{(1+0.5r\theta\sin^2\phi_i)}\right| \quad , \; 2\phi_i={}^{i\pi}/_{M}$$

therefore, the difference method (4.3.4) is stable if:

$$\begin{aligned} &\text{for } 0 \leq \theta < {}^1/_2, && {}^{\sigma k}/_{h^2} \leq {}^1/_{(2-4\theta)}, \\ &\text{for } {}^1/_2 \leq \theta \leq 1, && \text{no restriction.} \end{aligned} \qquad (4.3.8)$$

For the simple explicit method, (θ=0), (4.3.8) reduces to $r={}^{k\sigma}/_{h^2}\leq{}^1/_2$. The Crank-Nicolson method ($\theta={}^1/_2$) and the fully implicit method (θ=1) are unconditionally stable. The optimal method, $\theta={}^1/_2-{}^{h^2}/_{12k\sigma}$, which has local discretization error $O(k^2+h^4)$, is a member of the conditionally stable family, however, for this choice $(2-4\theta)^{-1}={}^{3k\sigma}/_{h^2}$ and so this method is stable.

Exercise 4.3.3 Show that the von Neumann analysis applied to the implicit difference method (4.3.4) produces the same stability condition as given by (4.3.8).

As with explicit methods we can improve the accuracy of implicit methods either by using extrapolation or by including more points of the mesh. With the former, Lawson and Morris (1978) and Gourlay and Morris (1980), have combined a scheme of the form (4.3.4) with a similar scheme using half step lengths, ${}^k/_2$, to produce accurate stable methods. However, Cash (1984) has shown that these methods are equivalent to applying implicit Runge-Kutta methods to the differential equation (4.2.4). More accurate schemes can also be obtained by applying the equivalent of the multi-step Adams methods to the differential equation ${}^{du}/_{dt}=\sigma Au$, but this can only improve the accuracy with respect to k. To improve the accuracy with respect to h we can either include more points at each time level, but this will cause problems near the boundaries at x=0 and x=1 as insufficient points will be available, or else use the equivalent of Numerov's method. For the latter we write the differential equation $u_t=\sigma u_{xx}$ as $u_{xx}=u_t/_\sigma$ which effectively gives a two point boundary value problem at each time level of the mesh $\Omega_{h,k}$. Applying Numerov's method gives

$$\frac{\delta_x^2 U_m^{n+1}}{h^2} = \frac{\{[U_{m+1}^{n+1}-U_{m+1}^{n}]/k + 10[U_m^{n+1}-U_m^{n}]/k + [U_{m-1}^{n+1}-U_{m-1}^{n}]/k\}}{12\sigma}. \quad (4.3.9)$$

Using a weighted average of u_{xx} at the nth and $n+1^{th}$ time levels gives the more accurate scheme

$$^1/_2\{\ \delta_x^2 U_m^{n+1} + \delta_x^2 U_m^{n}\ \} =$$

$$h^2/_{12k\sigma}\{[U_{m+1}^{n+1}-U_{m+1}^{n}] + 10[U_m^{n+1}-U_m^{n}] + [U_{m-1}^{n+1}-U_{m-1}^{n}]\}, \quad (4.3.10)$$

but this is nothing more than the general implicit method (4.3.4) with $\theta={}^1/_2 - {}^{h^2}/_{12k\sigma}$ and has been discussed previously.

Exercise 4.3.4 Show that there is a linear combination of the methods

(i) $$(1 + r\theta\ \delta_x^2)U_m^{n+1} = (1 + r(1-\theta)\ \delta_x^2)U_m^{n},$$

and

(ii) $$(1 + 0.5r\gamma\ \delta_x^2)U_m^{n+1/2} = (1 + 0.5r(1-\gamma)\ \delta_x^2)U_m^{n},$$

$$(1 + 0.5r\gamma\ \delta_x^2)U_m^{n+1} = (1 + 0.5r(1-\gamma)\ \delta_x^2)U_m^{n+1/2},$$

which gives a combined method which is 3rd order in time for linear parabolic equations. It is possible to show that no such formulae exist for non-linear equations whose order exceeds 2. (See Cash (1984)).

An alternative means of improving the spatial accuracy is given by the so called semi-discrete methods. As a simple example consider the equation $u_t=\sigma u_{xx}$, $0\leq x\leq 1$, $t\geq 0$, subject to $u(0,t)=u(1,t)=0$ and $u(x,0)=\phi(x)$, $0\leq x\leq 1$, for which a solution of the form

$$u_N(x,t) = \sum_{i=1..N} \alpha_i(t)\psi_i(x), \quad (4.3.11)$$

is sought. Here $\psi_i(x)$, $i=1,..,N$, form a given set of basis functions each of which satisfies the homogeneous boundary conditions $\psi_i(0)=\psi_i(1)=0$. The solution of the partial differential equation now becomes the problem of selecting functions $\alpha_i(t)$, $i=1,..,N$, such that the error $|u(x,t)-u_N(x,t)|$ is as small as possible for the given value of N. Substituting the expression (4.3.11) into the differential equation gives

$$\sum_{i=1..N} \alpha_i'(t)\psi_i(x) = \sigma \sum_{i=1..N} \alpha_i(t)\psi_i''(x).$$

We cannot expect to satisfy this equation exactly for all values of x in (0,1) but selecting N distinct points, $0\leq x_1<..<x_N\leq 1$, the Method of Collocation gives

$$\sum_{i=1..N} \alpha_i'(t)\psi_i(x_j) = \sigma \sum_{i=1..N} \alpha_i(t)\psi_i''(x_j), \quad j=1,..,N.$$

This gives a system of N ordinary differential equations which may be solved to find the coefficients $\alpha_i(t)$, i=1,..,N. Usually these equations are stiff in the sense discussed in Chapter 2, Section 11 and so an implicit method is normally required to solve them numerically. It is also possible to adopt a Galerkin or Rayleigh-Ritz type method to determine suitable expressions for the coefficients α_i. (See Chapter 3, Section 5.)

4.4 Generalisations

So far we have discussed the solution of the simple parabolic equation $u_t=\sigma u_{xx}$, $u(x,0)=\phi(x)$, $0\leq x\leq 1$, $u(0,t)=u(1,t)=0$, where σ was assumed to be constant. We now consider how such numerical methods can be extended and begin by generalising the boundary conditions.

Example 4.4.1 Compute a difference solution for the problem

$$u_t=u_{xx}, \; u(x,0)=\phi(x), \; 0\leq x\leq 1, \tag{4.4.1}$$

$$\partial u/\partial x+p(t)u=q(t) \text{ at } x=0, \; u(1,t)=s(t), \; t>0.$$

Solution. In order to demonstrate the method of solution we use the difference equation (4.2.2). The region $(0,1)\times(0,T)$ is covered by a mesh with spacings h and k, where $Mh=1$. The initial conditions give $U_m^0=\phi(mh)$, $m=0,1,..,M$. We then proceed as follows:-

(i) Apply the difference equation to obtain $\{U_m^1\}$, $m=1,..,M-1$.

(ii) At $x=1$ the second boundary condition gives $U_M^1=s(k)$. (We shall assume that $s(0)=\phi(1)$ for compatibility of the initial and boundary conditions at $x=1$, $t=0$.)

(iii) At $x=0$ we could replace the derivative ${}^{\partial u}/_{\partial x}$ by the difference $[U_1^1-U_0^1]/_h$ but this is only an $O(h)$ approximation which will detract from the overall $O(h^2)$ of the difference method. We overcome this problem by introducing a fictitious point at $x=-h$, as was done with the two point boundary value problem in Example 3.4.1, and then approximate the derivative ${}^{\partial u}/_{\partial x}$ by $[U_1^1-U_{-1}^1]/_{2h}$, where U_{-1}^1 is the value at the extended mesh point $x=-h$, $t=k$. Therefore, at $x=0$ the given boundary condition becomes

$$[U_1^1-U_{-1}n_{-1}^1]/_{2h} + p(k)U_0^1 = q(k),$$

or

$$U_{-1}^1=U_1^1-2h[q(k)-p(k)U_0^1]. \qquad (4.4.2)$$

From the difference equation (4.2.2), with $m=0$ and $n=0$, we obtain

$$U_0^1 = U_0^0 + r[U_1^0-2U_0^0+U_{-1}^0] \qquad (4.4.3)$$

and then eliminate the value U_{-1}^1 between equations (4.4.2) and (4.4.3) to give

$$U_0^1=(1-2r+2rhp(0))U_0^0 + 2rU_1^0 - 2rhq(0).$$

Repeating steps (i), (ii) and (iii) for $n=2,3,...$ determines values for $\{U_m{}^n\}$, $m=0,1,..,M-1$. This combination of difference scheme and approximation of the boundary condition at $x=0$ has an overall error $O(k+h^2)$ and is therefore consistent with the given problem. In order to show that it is convergent it is necessary to show that it is stable. Accordingly, we collect together the M equations which determine $U^{n+1}=[U_0{}^{n+1},U_1{}^{n+1},\ldots,U_{M-1}{}^{n+1}]^T$ in terms of $U^n=[U_0{}^n,U_1{}^n,\ldots,U_{M-1}{}^n]^T$ and write them in the form

$$U^{n+1} = BU^n + b, \tag{4.4.4}$$

where

$$B = \begin{bmatrix} 1-2r+2rhp((n-1)k) & 2r & 0 & \cdot & & 0 \\ r & 1-2r & r & 0 & \cdot & 0 \\ 0 & \cdot & \cdot & \cdot & \cdot & \cdot \\ \cdot & & \cdot & r & 1-2r & r \\ 0 & & & & r & 1-2r \end{bmatrix} \quad b = \begin{bmatrix} -2rhq((n-1)k) \\ 0 \\ \cdot \\ 0 \\ rs((n-1)k) \end{bmatrix}.$$

We emphasize at this point that the difference solution is not determined from equation (4.4.4) but pointwise as described previously. The difference method is stable if $\|B\|\leq 1$ but as there is no simple explicit form for the eigenvalues of B we apply Gerschgorin's Theorem to estimate their location. If $r\leq {}^1/_2$ and $p(x)<0$, $0\leq x\leq 1$, then all the eigenvalues of B lie within or on the unit circle and so the method is stable. If either of these conditions is violated we cannot conclude that the method is unstable, but merely that the result produced by Gerschgorin's Theorem is inconclusive.

Exercise 4.4.1 Show that provided $p(x)<0$ the Crank-Nicolson method applied to the problem (4.4.1) is unconditionally stable. Notice that since the matrix B is no longer constant it is necessary to carry out an LU decomposition at each step.

Example 4.4.1 demonstrates that the extension to derivative boundary conditions is straightforward; however, there is one problem introduced by this approach. It is possible to show that the solution of the equation $u_t=\sigma u_{xx}$, subject to the simple boundary conditions (4.1.3), is composed of two components, the first is a transient part which decays to zero and the second which although it depends on both x and t is usually referred to as the steady state solution. The same is true of the solution of the difference equation (4.3.4) when applied to this problem. However, Crank and

Parker (1964) have shown that if the differential equation is subject to derivative boundary conditions then the transient part of the difference solution decays to some small value ε which depends only upon the mesh size h for large values of n. Such a situation is called persistent discretization error and is unavoidable. A simple but complete example has been given by Smith (1978).

We now consider the solution of the equation $u_t=\sigma u_{xx}$, x and t in some region Ω, where $\sigma=\sigma(x,t)$ is a slowly varying function. The simple explicit method may still be applied to this problem, all that is necessary is to replace σ in the recurrence relation (4.2.2) by $\sigma_m{}^n$. This gives a method which has discretization error $O(k+h^2)$ and is therefore consistent. We can no longer determine the eigenvalues of the matrix (4.2.3) directly but we can apply Gerschgorin's Theorem from which we obtain a stability condition that the mesh sizes must satisfy

$$(k\sigma_{max})/h^2 \leqslant 1/2,$$

where $\sigma_{max}=\max(\sigma(x,t))$, for x and t in Ω. For each of the other explicit methods a similar result applies, i.e. we simply replace σ by $\sigma_m{}^n$ and compute the difference solution in the same way. The application of Douglas's method is not strictly valid for it depends on replacing $\partial^2u/\partial t^2$ by $\sigma^2\,\partial^4u/\partial x^4$ which is only possible if σ is independent of both x and t. This can be overcome by a suitable modification of the difference equations but if σ does not change too rapidly Douglas's method can be applied unchanged. For the implicit methods it is necessary to replace σ not by $\sigma_m{}^n$ but by $\sigma_m{}^{n+1/2}$ since $[U_m{}^{n+1}-U_m{}^n]/k$ gives an $O(k^2)$ approximation for u_t at the point $(x,t+k/2)$. The optimal scheme with $\theta=1/2-h^2/12k\sigma$ is again not strictly applicable but a simple modification of (4.3.9) can be used. Again if σ does not change too rapidly it is possible to use $\theta=1/2-h^2/12k\sigma_m{}^n$.

Finally in this section, as an example of the extension of finite difference methods, let us consider the solution of the inhomogeneous linear differential equation

$$v_t = v_{xx} + \beta v, \qquad (4.4.5)$$

subject to

$$v(0,t)=v(1,t)=0 \text{ and } v(x,0)=\phi(x),\ 0\leqslant x\leqslant 1.$$

Thus far we have assumed that the solution of the given differential equation is bounded; for this problem this is not necessarily the case. The equation (4.4.5) can be written in the form

$$\partial/\partial t(e^{-\beta t}v) = \exp^{-\beta t}v_{xx},$$

the solution of which is given by $v(x,t)=e^{\beta t}u(x,t)$, where u satisfies the equation $u_t=\sigma u_{xx}$. If $\beta<0$ then the von Neumann method gives a stability criterion that the mesh sizes should satisfy $r\leqslant 2/(4-\beta h)$. However, if β is positive and sufficiently large the solution of equation (4.4.5) will grow, and to demand that the difference solution remains bounded is too severe a restriction. In order to allow for the possibility of a growing solution we require that when applying the matrix stability analysis all eigenvalues are bounded by a term of the form $1+O(k)$. For the von Neumann method we require that the amplification factor, ρ, should be bounded by a similar term. This will allow the difference solution to grow; but not too quickly. A full justification lies beyond the scope of this text but details are given by Richtmyer and Morton (1967). Richtmyer and Morton also show that the effect of lower order terms is of little practical importance when considering the stability of a difference method.

Exercise 4.4.2 Show that the simple explicit method

$$V_m^{n+1} = V_m^n + k/h^2[V_{m+1}^n-2V_m^n+V_{m-1}^n] + k\beta V_m^n,$$
$$0 < m < M,\ Mh=1,\ n > 0,$$

applied to the differential equation (4.4.5) with $\beta<0$ is stable provided $r \leqslant 2/(4-\beta h^2)$.

Exercise 4.4.3 The differential equation

$$u_t = u_{xx} + au_x + bu, \qquad (4.4.6)$$

can be approximated by the difference equation

$$\frac{[U_m^{n+1}-U_m^n]}{k} = \frac{\delta_x^2 U_m^n}{h^2} + a\frac{[U_{m+1}^n-U_{m-1}^n]}{2h} + bU_m^n. \qquad (4.4.7)$$

Show that this is a consistent approximation and determine the leading terms in the discretization error. Determine the range of values of k and h for which (4.4.7) is stable. If $a=a(x,t)$ how would this alter the solution of this problem?

Exercise 4.4.4 As an alternative to the difference equation (4.4.7) the derivative u_x, in equation (4.4.6), could be replaced by either

(i) $[U_m^n-U_{m-1}^n]/h$

or

(ii) $[U_{m+1}^n-U_m^n]/h$.

How does this alter the order and stability of the resulting method?

4.5 An Introduction to Non-linear Equations

A non-linear parabolic partial differential equation, written in the form

$$u_t=f(u_{xx},u_x,u,x,t), \qquad (4.5.1)$$

subject to suitable initial and boundary conditions, is properly posed as an initial value problem provided $\partial f/\partial u_{xx}>0$. In order to construct a difference method we replace each derivative by a suitable difference approximation and then proceed as before. It is necessary to consider the consistency and stability of the resulting difference method, both of which are clearly required to ensure that the computed difference solution converges to the solution of the differential equation as the finite difference mesh is refined. Lax's Equivalence Theorem has been extended by Rosinger (1980) to show that for a properly posed initial value problem and a consistent difference approximation, stability and convergence are equivalent.

To establish consistency we examine the local discretization error as with linear equations: stability is a little more difficult as we cannot use a matrix stability analysis nor apply a global von Neumann method. However, provided the solution does not change too rapidly we can still apply the von Neumann method in a local way.

Example 4.5.1 To demonstrate a local application of the von Neumann method let us consider the solution of the equation $u_t=(u^5)_{xx}=\partial/\partial x(5u^4\partial u/\partial x)$. Comparing this equation with (4.1.1) we have $\sigma=5u^4$. The general single step method (4.3.4) applied to the equation $u_t=\sigma u_{xx}$, where σ is a constant, is unconditionally stable for $1/2\leqslant\theta\leqslant1$ and stable provided $\sigma k/h^2<1/(2-4\theta)$ when $0\leqslant\theta<1/2$. The corresponding condition when applied to the given non-linear equation is that the method is stable provided

$$5u^4k/h^2 < 1/(2-4\theta)$$

when $0\leqslant\theta<1/2$ and unconditionally stable otherwise. Therefore, as the solution is computed it is necessary to check whether this condition is satisfied; if not then the step sizes should be adjusted accordingly.

In some circumstances it is possible to show that the solution is bounded, and hence stable, by more indirect means. For example, Hagan (1981) has shown that only monotonic wave solutions of equation (4.5.1) are stable; in the sense that they are insensitive to small perturbations. This property can be used to show that the corresponding difference equation is also monotonic and hence bounded.

Example 4.5.2 Consider the following problem due to Fisher.

$$u_t = u_{xx} + u(1-u), \ -\infty<x<\infty, \ t\geq 0, \ u(-\infty,t)=1, \ u(\infty,t)=0.$$

This problem has a monotonic wave solution. It is straightforward to show that if the simple explicit difference method is applied to this problem it produces a monotonic solution provided $k\leq h^2/(2-h^2)$. As the difference solution is monotonic and constrained, by the boundary conditions, to lie in the interval (0,1) it is bounded, hence, the method is stable. (See Murray (1976), Quinney (1979)).

With linear partial differential equations it clear that the unconditional stability and additional accuracy of implicit methods are very attractive. However, with non-linear equations it is necessary to solve a system of non-linear equations at each step in order to advance the difference solution through the mesh. Although such equations can be solved by a simple iteration the additional work that this involves makes a simple explicit method more effective in many cases. There is no clear answer as to whether an explicit or implicit method should be used, it depends on the accuracy of the solution which is required and the facilities available.

4.6 Equations Involving More Than One Spatial Variable

In the introduction to this chapter it was shown that the temperature within a uniform block of material occupying region of space D_0 satisfies the equation $u_t=\sigma\nabla^2 u$, subject to suitable initial and boundary conditions. In one spatial dimension this reduces to the equation (4.1.2); similarly in two dimensions it becomes

$$u_t = \sigma(u_{xx} + u_{yy}), \quad (x,y) \text{ in } D_0, \ t\geq 0. \qquad (4.6.1)$$

In this section we introduce difference methods for the solution of equation (4.6.1) subject to the initial condition, $u(x,y,0)=\phi(x,y)$, and the boundary condition $u(x,y,t)=g(x,y,t)$ for (x,y) in ∂D_0, the boundary of D_0. For simplicity we consider only the simple region $D_0=(0,1)\times(0,1)$. The region $\Omega=D_0\times(0,T)$ is covered with a uniform mesh $\Omega_{h,k}=\{(x,y,t) : x=mh,\ y=\ell h,\ 0<m,\ell<M,\ Mh=1,\ t=nk,\ n\geq 0\}$ and the points where the mesh $\Omega_{h,k}$ intersect the boundary ∂D_0 are denoted by $\partial\Omega_{h,k}$. Let $U_{m\ell}{}^n$ be a computed approximation for $u(mh,\ell h,nk)$ then the simplest difference approximation for (4.6.1) is given by

$$U_{m\ell}{}^{n+1} = U_{m\ell}{}^n + r[\delta_x^2 U_{m\ell}{}^n + \delta_y^2 U_{m\ell}{}^n] = (1 + r(\delta_x^2+\delta_y^2))U_{m\ell}{}^n, \qquad (4.6.2)$$

where δ_x and δ_y denote central difference operators with respect to x and y respectively. By expanding in a Taylor series about the point $(mh,\ell h,nk)$ it is straightforward to show that this method has a local discretization error which is proportional to $O(k+h^2)$ and so the method is consistent with the differential equation (4.6.1). The stability of the scheme (4.6.2) may be examined by considering a difference solution which is of the form $U_{m\ell}{}^n=A\rho^n\exp(im\phi)\exp(i\ell\psi)$. Substituting this expression into (4.6.2) the amplification factor, ρ, is given by

$$\rho=1 - 4r(\sin^2\phi/2 + \sin^2\psi/2), \quad r=k\sigma/h^2.$$

Therefore, for stability we require that $0<r\leq 1/4$.

Exercise 4.6.1 An alternative to (4.6.2) is given by

$$U_{m\ell}{}^{n+1}= (1 + r\delta_x^2)(1 + r\delta_y^2)U_{m\ell}{}^n.$$

Show that this method has a local discretization error proportional to $O(k+h^2)$. In addition show that this method is stable provided $0<r\leq 1/2$. If $r=1/6$ show that this method has local error proportional to $O(k^2+h^4)$.

Although it is possible to construct unconditionally stable explicit methods which correspond to the DuFort and Frankel scheme, implicit methods are to be preferred. For example, the

Crank-Nicolson method for equation (4.6.1) is given by

$$U_{m\ell}^{n+1} = U_{m\ell}^{n} + 0.5r[\delta_x^2(U_{m\ell}^{n+1}+U_{m\ell}^{n})+\delta_y^2(U_{m\ell}^{n+1}+U_{m\ell}^{n})], \quad (4.6.3)$$

This method will involve the solution of $(M-1)^2$ linear equations at each time level.

Exercise 4.6.2 Show that the method (4.6.3) is $O(k^2+h^2)$ and unconditionally stable.

Exercise 4.6.3 The Crank-Nicolson method may also be written

$$(1-0.5r\delta_x^2)(1-0.5r\delta_y^2)U_{m\ell}^{n+1} = (1+0.5r\delta_x^2)(1+0.5r\delta_y^2)U_{m\ell}^{n}. \quad (4.6.4)$$

Show that this is computationally equivalent to (4.6.3).

Exercise 4.6.4 The Mitchell-Fairweather formula is given by

$$(1-\alpha\delta_x^2)(1-\alpha\delta_y^2)U_{m\ell}^{n+1} = (1+\beta\delta_x^2)(1+\beta\delta_y^2)U_{m\ell}^{n}$$

where $\alpha={}^r/_2-{}^1/_{12}$ and $\beta={}^r/_2+{}^1/_{12}$. Show that this method is $O(k^2+h^4)$. Notice the similarity between this scheme and the method due to Douglas given in Example 4.2.2.

The need to solve large systems of equations at each step is a considerable drawback when applying methods which are implicit in both the x and y directions. As an alternative we consider methods which are constructed from two alternating steps. The first step applies an implicit method in the x-direction and an explicit method in the y-direction to produce an intermediate solution; the second step produces the solution at the $n+1^{th}$ level of the mesh by interchanging the implicit and explicit procedures. This results in solving $2(M-1)$ sets of $(M-1)$ tridiagonal equations in order to produce $\{U_{m\ell}^{n+1}\}$, $0<m,\ell<M-1$, from $\{U_{m\ell}^{n}\}$, $0\leqslant m,\ell\leqslant M$. Due to the method of their construction such methods are called **Alternating Direction Implicit** (ADI) methods or ADI splits.

Example 4.6.1 (a) The Peaceman-Rachford ADI method is given by

$$\begin{aligned} &\text{(i) } V_{m\ell}^{n+1/2} = U_{m\ell}^{n} + 0.5r\,[\delta_x^2 V_{m\ell}^{n+1/2} + \delta_y^2 U_{m\ell}^{n}], \\ &\text{(ii) } U_{m\ell}^{n+1} = V_{m\ell}^{n+1/2} + 0.5r\,[\delta_x^2 V_{m\ell}^{n+1/2} + \delta_y^2 U_{m\ell}^{n+1}]. \end{aligned} \tag{4.6.5}$$

These equations can also be written as

$$\begin{aligned} &\text{(i) } (1 - 0.5r\delta_x^2)V_{m\ell}^{n+1/2} = (1 + 0.5r\,\delta_y^2)U_{m\ell}^{n}, \\ &\text{(ii) } (1 - 0.5r\,\delta_y^2)U_{m\ell}^{n+1} = (1 + 0.5r\,\delta_y^2)V_{m\ell}^{n+1/2}. \end{aligned}$$

Eliminating the intermediate solution $V_{m\ell}^{n+1/2}$ reduces this method to the Crank-Nicolson formula (4.6.4).

(b) The Douglas-Rachford Split is given by

$$\begin{aligned} &\text{(i) } V_{m\ell}^{n+1/2} = U_{m\ell}^{n} + 0.5r\delta_x^2[U_{m\ell}^{n+1}+U_{m\ell}^{n}] + r\delta_y^2 U_{m\ell}^{n}, \\ &\text{(ii) } U_{m\ell}^{n+1} = U_{m\ell}^{n}+r\delta_x^2[V_{m\ell}^{n+1/2}+U_{m\ell}^{n}] + 0.5r\,[U_{m\ell}^{n+1}+U_{m\ell}^{n}], \end{aligned} \tag{4.6.6}$$

which is also equivalent to (4.6.4).

Exercise 4.6.5 One of the many other possible ADI methods is given by the following

$$\begin{aligned} &\text{(i) } (1 - 0.5r\delta_x^2)V_{m\ell}^{n+1/2} = (1 + 0.5r\delta_x^2)(1+0.5r\delta_y^2)U_{m\ell}^{n}, \\ &\text{(ii) } (1 - 0.5r\delta_y^2)U_{m\ell}^{n+1} = V_{m\ell}^{n+1/2}, \end{aligned} \tag{4.6.7}$$

which is due to D'Yakanov. Show that this method is consistent of order k^2+h^2 with the differential equation (4.6.1) and that it is unconditionally stable.

With each of these ADI operators it is necessary that the computed solution satisfy all the initial and boundary conditions. Clearly, where the mesh $\Omega_{h,k}$ meets the boundary ∂D_0 we set

$U_{m\ell}{}^{n+1}=g(mh,\ell h,nk)$. At the intermediate steps it is tempting to associate $V_{m\ell}{}^{n+1/2}$ with $u(mh,\ell h,(n+1/2)k)$ and, therefore, put $V_{m\ell}{}^{n+1/2}=g(mh,\ell k,(n+1/2)k)$; but this is not necessarily the correct boundary condition. For example, eliminating $\delta_x^2 V_{m\ell}{}^{n+1/2}$ between (4.6.5) (i) and (4.6.5) (ii) gives

$$V_{m\ell}{}^{n+1/2} = 1/2(U_{m\ell}{}^{n+1}+U_{m\ell}{}^{n}) - 0.25r\ \delta_y^2(U_{m\ell}{}^{n+1}-U_{m\ell}{}^{n})$$

and so the correct boundary condition is given by

$$V_{m\ell}{}^{n+1/2} = 0.5(g_{m\ell}{}^{n+1}+g_{m\ell}{}^{n}) - 0.25r\ \delta_y^2(g_{m\ell}{}^{n+1}-g_{m\ell}{}^{n}), \qquad (mh,\ell h) \text{ in } \partial\Omega_{h,k},$$

where $g_{m\ell}{}^{n}=g(mh,\ell h,nk)$, which only reduces to $g_{m\ell}{}^{n+1/2}$ if g is independent of t.

Exercise 4.6.6 Show that the correct boundary conditions for the Douglas-Rachford ADI method are given by

$$\text{(i)}\ V_{m\ell}{}^{n+1/2} = (1 - 0.5r\delta_y^2)g_{m\ell}{}^{n+1} + 0.25r\ \delta_y^2 g_{m\ell}{}^{n},$$

$$\text{(ii)}\ U_{m\ell}{}^{n+1} = g_{m\ell}{}^{n+1}, \qquad (mh,\ell h) \text{ in } \partial\Omega_{h,k}.$$

Exercise 4.6.7 Show that the correct boundary conditions for the D'Yakanov split ADI are given by

$$\left.\begin{array}{ll}\text{(i)} & V_{m\ell}{}^{n+1/2} = (1-0.5r\delta_y^2)g_{m\ell}{}^{n+1},\\ \text{(ii)} & U_{m\ell}{}^{n+1} = g_{m\ell}{}^{n+1}.\end{array}\right. \qquad (mh,\ell h) \text{ in } \partial\Omega_{h,k}$$

Further details and other alternative ADI methods are given in a survey article in Jacobs (1977). The extension of such difference methods to problems involving more complicated geometry, non-constant coefficients, additional spatial variables and certain non-linearities is relatively straightforward. The analysis of such methods, i.e. an examination of the consistency and stability, can be considered in analogy with Sections 4.5 and 4.6. For examples of such generalisations see Ames (1977), Jain (1978) and Richtmyer and Morton (1967).

CHAPTER 5

Hyperbolic Partial Differential Equations

5.1 Introduction

The wave equation $\partial^2 u/\partial t^2 = c^2\ \partial^2 u/\partial x^2$ is an example of a linear second order partial differential equation and can be classified as hyperbolic. (See Appendix 2). This equation, subject to suitable initial and boundary conditions can be used as a model for numerous physical situations.

Example 5.1.1 Consider the motion of a tightly stretched string which vibrates in the x-y plane and is fixed at (0,0) and (0,L). At some instant of time let us suppose that the shape of the string is as given by Figure 5.1.1.

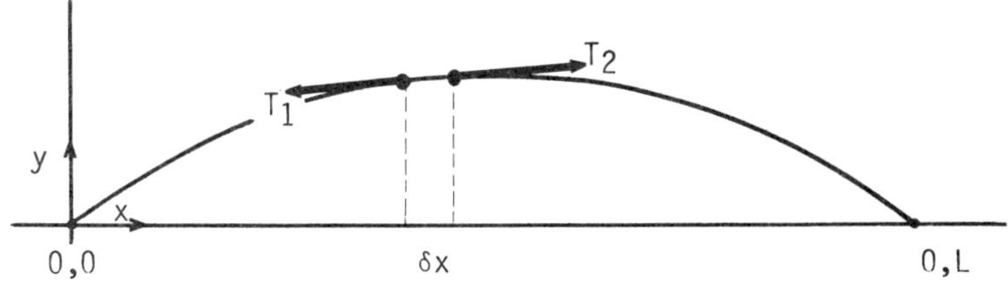

Figure 5.1.1 The displacement of the stretched string at time t.

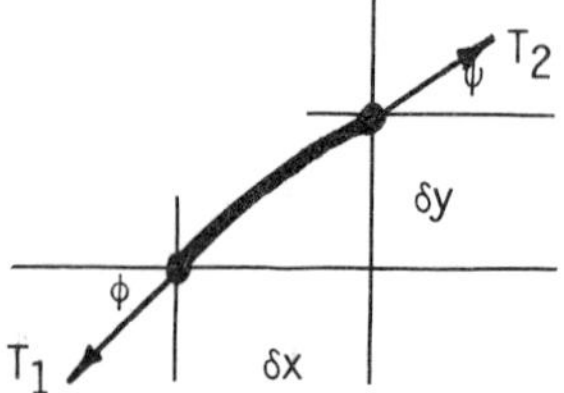

Figure 5.1.2 The forces acting on a segment δx.

Let δx be a small segment of the string which has density ρ per unit length then by Newton's second law of motion

$$\rho\delta x\,\partial^2 y/\partial t^2 = \text{Net force acting on segment} = T_2\sin\psi - T_1\sin\phi.$$

Resolving the forces acting on the segment parallel to the x-axis gives $T_2\cos\psi = T_1\cos\phi$ and setting $\psi=\phi+\delta\phi$ then $T_2\cos(\phi+\delta\phi)=T_1\cos\phi$. If $\delta\phi$ is sufficiently small then $T_2 \simeq T_1 = T$ which substituted into the equation of motion produces

$$\partial^2 y/\partial t^2 = T/\rho\;(\sin(\phi+\delta\phi)-\sin(\phi))/\delta x.$$

If the displacement of the string is small then $\sin\phi \simeq \tan\phi \simeq \partial y/\partial x$ and as $\delta x \to 0$ this equation becomes

$$\partial^2 y/\partial t^2 = T/\rho\;\partial^2 y/\partial x^2, \quad 0 \leqslant x \leqslant L,\; t \geqslant 0. \qquad (5.1.1)$$

At the points (0,0) and (0,L) the ends of the string are fixed; therefore, appropriate boundary conditions for this problem are given by $y(0,t)=y(L,t)=0$. Any initial conditions will depend upon the shape of the string and its velocity at the initial time $t=0$, for example, if the string is plucked by pulling its centre a distance ε from its equilibrium position and releasing it from rest then

$$y(x,0) = \begin{cases} 2\varepsilon/L, & 0 \leqslant x \leqslant L/2, \\ 2\varepsilon/L(L-x), & L/2 < x \leqslant L, \end{cases} \qquad \partial y/\partial t(x,0)=0,\; 0 \leqslant x \leqslant L, \qquad (5.1.2)$$

Equation (5.1.1) can be written as a system of first order equations by setting $u = \partial y/\partial t$ and $v = \partial y/\partial x$ which produces

$$\begin{aligned} u_t - cv_x &= 0, \\ v_t - cu_x &= 0, \end{aligned} \qquad (5.1.3)$$

where $c^2 = T/\rho$. These equations can then be rewritten as the system of equations

$$\mathbf{U}_t + A\,U_x = 0, \qquad (5.1.4)$$

where $U = (u,v)^T$ and $A = \begin{bmatrix} 0 & -c \\ -c & 0 \end{bmatrix}$

In order to construct a difference approximation for the solution of the problem given by equations (5.1.1) and (5.1.2), or equivalently (5.1.4) subject to suitable initial and boundary conditions, we begin by examining the single scalar equation

$$u_t + cu_x = 0, \quad c>0, \ x,t \geq 0. \qquad (5.1.5)$$

let us suppose that the solution of this equation is known along a curve C in the x-t plane and consider how the solution at a point P, (x,t), is related to the solution at a neighbouring point, Q, (x+h,t+k), which lies on another curve Γ intersecting C at P.

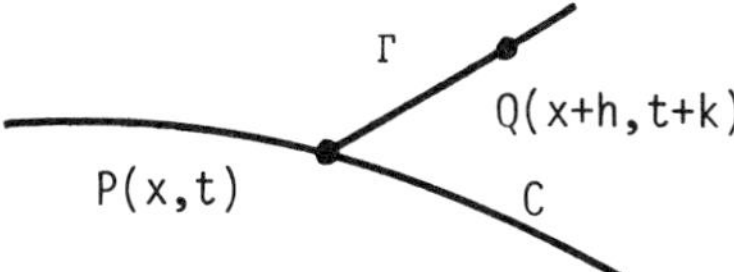

Applying the chain rule the derivative of u in the direction Γ is given by

$$D(u) = \partial u/\partial t \; dt + \partial u/\partial x \; dx \qquad (5.1.6)$$

and combining equation (5.1.5) and (5.1.6) gives

$$\begin{bmatrix} c & 1 \\ dx & dt \end{bmatrix} \begin{bmatrix} u_x \\ u_t \end{bmatrix} = \begin{bmatrix} 0 \\ D(u) \end{bmatrix}. \qquad (5.1.7)$$

These equations determine u_x and u_t uniquely except when $cdt-dx=0$, i.e. except along the characteristic lines $x-ct=$constant. (See Appendix 2). Along a characteristic line there can be no solution of equation (5.1.7) unless

$$\text{Det}\begin{bmatrix} c & 0 \\ dx & D(u) \end{bmatrix} = 0, \qquad (5.1.8)$$

that is $D(u)=0$, and then the solution of problem (5.1.5) is constant along its characteristic curves. Therefore, the general solution of equation (5.1.5) is given by $u(x,t)=f(x-ct)$, where $f(z)$ is any arbitrary function of the single variable z. Since the function f, and hence u, takes the same value at every point along a characteristic the solution of this equation at any point is determined by where the characteristic passing through this point meets any initial or boundary conditions.

Example 5.1.2 Determine the characteristic lines for the equation $u_t-2u_x=0$. Find the solution of this equation subject to $u(x,0)=1$, $x \geqslant 0$ and $u(0,t)=0$, $t \geqslant 0$.

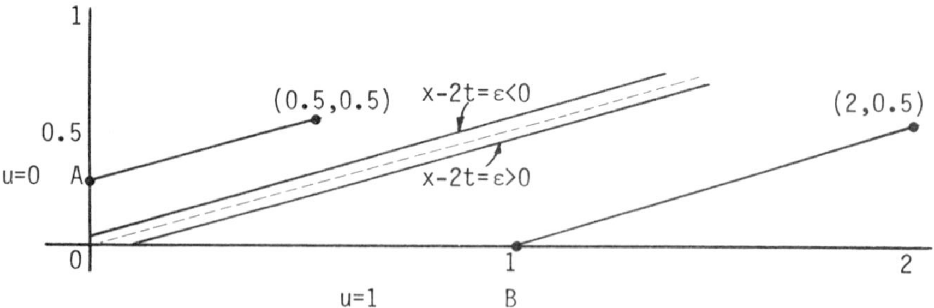

Figure 5.1.3 Characteristic lines of the equation $u_t-2u_x=0$.

The solution at $x=0.5$, $t=0.5$ is given by tracing the characteristic through this point to where it meets the boundary of the region,

$x \geq 0$, $t \geq 0$, i.e. the point A=(0,0.25), from which we conclude that $u(0.5,0.5)=u(0,0.25)=0$. For the point (2,0.5) the characteristic passes through the x-axis at the point B and so $u(2,0.5)=1$. Notice that at each point along the line $x-2t=\varepsilon$, $\varepsilon>0$, $u(x,t)=1$ but if $\varepsilon<0$ the $u(x,t)=0$. Letting $\varepsilon \to 0$ the solution is clearly discontinuous across the characteristic $x-2t=0$. This demonstrates the fundamental nature of the characteristic curves and also that any discontinuity in the initial and/or boundary conditions associated with a hyperbolic partial differential equation will propagate in this way. Any useful numerical method for obtaining a solution of such an equation must exhibit the same behaviour.

Example 5.1.3 Find the general solution of $u_t+(1+x)u_x=0$.

Solution. From (5.1.6) the characteristic directions are given by

$$\text{Det}\begin{bmatrix} (1+x) & 1 \\ dx & dt \end{bmatrix} = 0, \quad \text{or } (1+x)dt - dx = 0$$

and so the characteristics themselves are given by the curves $t=\log(1+x)+C$. Any solution is constant along these lines and so the general solution is given by $u(x,t)=f(t-\log(1+x))$, where f is any arbitrary function.

Exercise 5.1.1 Show that the solution of $u_t+(1+x)u_x=0$ subject to $u(0,t)=e^{-t}$, $t \geq 0$, $u(x,0)=1+x$, $x \geq 0$, is given by

$$u(x,t) = \begin{cases} (1+x)e^{-t}, & t > \log(1+x), \\ (1+x)^2e^{-2t}, & t < \log(1+x). \end{cases}$$

Notice that even though u is continuous across $t=\log(1+x)$ neither $\partial u/\partial t$ not $\partial u/\partial x$ are.

Exercise 5.1.2 Determine the characteristics of the non-linear equation

$$tu_t + uu_x = -u^2$$

and hence obtain the solution of this equation which satisfies u=1 when t=0 and $0 \leqslant x \leqslant 1$. Determine the values (x,t) for which this solution is valid.

Further details regarding characteristics and hyperbolic equations can be found in texts such as Lefschetz (1968), Garabedian (1964) and Weinberger (1965).

5.2 Simple Difference Methods for First Order Hyperbolic Equations

We begin by considering the solution of the hyperbolic equation

$$u_t + cu_x = 0, \qquad (5.2.1)$$

where c is a positive constant, subject to the initial condition $u(x,0)=\phi(x)$, $x \geqslant 0$, and the boundary condition $u(0,t)=\psi(t)$, $t \geqslant 0$. The quarter plane $\Gamma=\{(x,t) : x \geqslant 0, t \geqslant 0\}$, is covered by a mesh which has spacing h in the x-direction and k in the t-direction and is denoted by $\Gamma_{h,k}$.

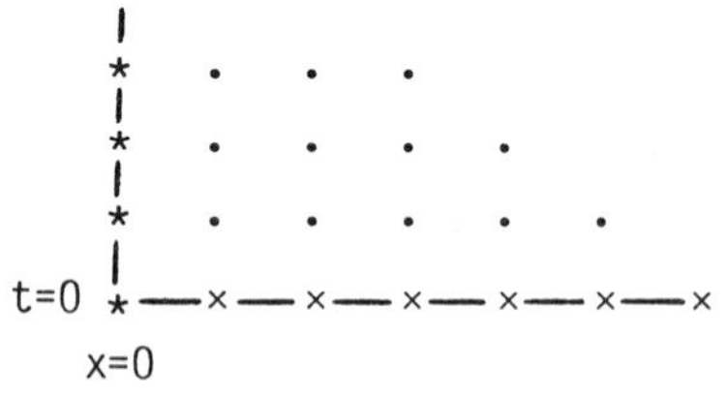

Figure 5.2.1 **The Covering Mesh $\Gamma_{h,k}$ for Γ.**

At each point, (mh,nk), of the mesh $\Gamma_{h,k}$ a value U_m^n is determined as an approximation for the solution u(mh,nk). The initial condition supplies values for $U_m^0=u(mh,0)=\phi(mh)$, m=0,1,...., and the boundary conditions give $U_0^n=u(0,nk)=\psi(nk)$, n=1,.... . (These points are denoted by "×" and "*" respectively in Figure 5.2.1.). At each internal point of the mesh the differential equation is replaced by a consistent difference approximation. For example, replacing u_t by $[U_m^{n+1}-U_m^n]/_k$ and u_x by $[U_{m+1}^n-U_m^n]/_h$ produces the difference approximation

$$[U_m^{n+1} - U_m^n]/_k + c[U_{m+1}^n - U_m^n]/_h = 0$$

or

$$U_m^{n+1} = U_m^n - {}^{kc}/_h [U_{m+1}^n - U_m^n], \quad (mh,nk) \text{ in } \Gamma_{h,k}. \qquad (5.2.2)$$

This difference method may then be used to determine values U_m^n, n>0,m>0. Notice that in order to determine U_m^{n+1} we need to know U_{m+1}^n and U_m^n, and to find these values we need U_{m+2}^{n-1}, U_{m+1}^{n-1} and U_m^{n-1} and so on. The set of values which is required to determine U_m^{n+1} is called its **Domain of Dependence.** For equation (5.2.2) it is given as follows:-

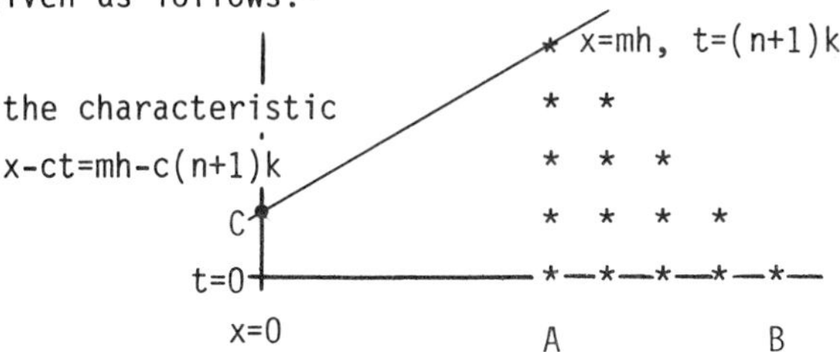

Figure 5.2.2 The Domain of Dependence for Equation (5.2.2).

As h and k tend to zero the shape of the domain of dependence remains the same; the points within it merely approach each other as the mesh is refined. Let us suppose that the domain of dependence of the point x=mh,t=(n+1)k meets the boundary of the region Ω at the line segment AB; accordingly, as the mesh is refined only the initial values between A and B will influence the solution at this point. The domain of dependence is said to be the domain of influence of the line segment AB. However, for this example the characteristic through mh,(n+1)k has a positive slope and meets the boundary at the point C, therefore, the behaviour of the boundary condition at C completely determines the solution of the differential equation at mh,(n+1)k. On the other hand, the solution of the difference equation at this point is determined by the behaviour of the initial condition between A and B. Clearly this is contradictory and so we might expect that the difference solution does not converge to the solution of the differential equation as the mesh is refined. The criterion that the characteristic through a point must lie within the domain of dependence of the same point is called the **Courant-Friedrichs-Lewy** (**CFL**) condition and is a necessary, but not sufficient, condition for the convergence of any difference method to a solution of a given hyperbolic equation. The difference method (5.2.2) is consistent with the differential equation (5.2.1) but it cannot produce a convergent difference solution. The reason for this can be found by examining the stability of the method and recalling that for a properly posed initial value problem stability is a necessary and sufficient condition for convergence. Let us consider possible solutions of (5.2.2) which are of the form $U_m^n = A\rho^n e^{im\phi}$. Substituting this expression into (5.2.2) the amplification factor, ρ, is given by

$$\rho = 1 - pc(e^{i\phi} - 1) = 1+pc(1-\cos\phi) - i\ pc\ \sin\phi,$$

where $p={}^k/_h$. We conclude that $|\rho|>1$, except when $\phi=0$, and so (5.2.2) is unstable.

Exercise 5.2.1 Show that if $c<0$ the difference method (5.2.2) can satisfy the CFL criterion and is stable provided $0<{}^k/_h|c|\leq 1$.

Exercise 5.2.2 By expanding in a Taylor series about the point $x=mh$, $t=nk$ show that the discretization error of the difference approximation for equation (5.2.1) given by

$$[U_m^{n+1} - U_m^n]/k + c\ [U_m^n - U_{m-1}^n]/h = 0, \qquad (5.2.3)$$

is $E(h,k)={}^k/_2\, u_{tt} + {}^h/_2\, u_{xx}$ + H.O.T. Show that this method satisfies the CFL condition provided that $0<{}^k/_h\leq{}^1/_c$, i.e. that the slope of the mesh is less than the slope of the characteristics.

We now consider implicit difference methods for the solution of the differential equation $u_t+cu_x=0$, where c is a positive constant.

Example 5.2.1 The difference approximation

$$[U_m^{n+1} - U_m^n]/k + c[U_m^{n+1} - U_{m-1}^{n+1}]/h = 0, \quad (mh,nk) \text{ in } \Gamma_{h,k}$$

is consistent, of O(h+k), with the differential equation (5.2.1). This equation may be written as

$$U_m^{n+1} (1 + pc) - pcU_{m-1}^{n+1} = U_m^n, \quad (5.2.4)$$

where $p={}^k/_h$, which demonstrates the implicit nature of this method. (More than one point at the n+1th level of the mesh). The domain of dependence of this method is given by all the points to the left and below x=mh, t=(n+1)k, therefore, any characteristic which has positive slope (c>0) will lie within this region and the CFL criterion is satisfied. Furthermore, $U_m^n=A\rho^n e^{im\phi}$ will be a solution of (5.2.4) provided

$$\rho = (1 + pc(1 - e^{im\phi}))^{-1}.$$

Such a solution is bounded if $|\rho|\leqslant 1$. This condition is satisfied for all mesh sizes, h and k, and so (5.2.4) is unconditionally stable.

Exercise 5.2.3 Show that the approximation

$$[U_m^{n+1} - U_m^n]/k + c[U_{m+1}^{n+1} - U_m^{n+1}]/h = 0, \ (mh,nk) \text{ in } \Gamma_{h,k},$$

is consistent with the equation $u_t+cu_x=0$ but does not satisfy the CFL criterion and is unconditionally unstable.

The method (5.2.4) is, by definition, implicit but it is explicit in practice. At m=0, a value for U_1^{n+1} can be found using U_0^{n+1}, the given boundary condition at x=0, and U_1^{n+1}. We now examine the stability of this method using the matrix stability analysis introduced in Chapter 4.

Example 5.2.2 Consider the differential equation $u_t+cu_x=0$, $0<x<1$, $t>0$. The strip $\Gamma=\{(x,t): 0<x<1, t>0\}$ is covered by a mesh with $Mh=1$ then if we set $U^n=[U_0{}^n,U_1{}^n,..,U_M{}^n]^T$ equation (5.2.4) can be written as

$$U^{n+1} = AU_n + b^n, \tag{5.2.5}$$

where

$$\mathbf{A} = \begin{bmatrix} 1-pc & 0 & & & & 0 \\ pc & 1-pc & 0 & & & 0 \\ 0 & pc & 1-pc & 0 & & 0 \\ & & \cdot & \cdot & \cdot & \cdot \\ 0 & & & 0 & pc & 1-pc \end{bmatrix} \qquad \mathbf{b}_n = \begin{bmatrix} pcU_0{}^n \\ 0 \\ \cdot \\ \cdot \\ 0 \end{bmatrix}$$

and $p={}^k/_h$. In order to show that this method is stable it is sufficient to show that $\|A\|\leq 1$ and as A is not symmetric we examine the eigenvalues of $A^T\mathbf{A}$. The product of A with its transpose is a tridiagonal matrix of the form

$$\begin{bmatrix} (1-pc)^2+(pc)^2 & pc(1-pc) & 0 & & & 0 \\ pc(1-pc) & (1-pc)^2+(pc)^2 & pc(1-pc) & 0 & & 0 \\ 0 & pc(1-pc) & (1-pc)^2+(pc)^2 & pc(1-pc) & 0 & 0 \\ & \cdot & \cdot & \cdot & & \\ & & \cdot & \cdot & \cdot & \\ 0 & & & 0 & pc(1-pc) & (1-pc)^2+(pc)^2 \end{bmatrix}$$

which has eigenvalues $(1-pc)^2+(pc)^2+2pc(1-pc)\cos{}^{j\pi}/_{(M+1)}$, $j=1,..,M$ and so $\|A\|\leq 1$ if $0<pc\leq 1$, which is the same condition obtained in Example 5.2.1.

All the methods obtained so far have been first order in h and k; we now consider more accurate methods. Previously difference methods were obtained by simply replacing any derivative by a consistent difference approximation; we now consider an alternative procedure. Let us assume that the solution of the differential equation at the point $x=mh$, $t=(n+1)k$, i.e. $u(x,t+k)$, can be expanded in a Taylor series about the point x,t to give

$$u(x,t+k) = u(x,t) + ku_t + {}^{k^2}/_2\, u_{tt} + \ldots. \qquad (5.2.6)$$

Writing the differential equation (5.2.1) as $u_t=-cu_x$ then

$$u(x,t+k) = u(x,t) - kcu_x + O(k^2).$$

Replacing u_x by the difference approximation $[u(x+h,t)-u(x-h,t)]/_{2h}$ gives

$$u(x,t+k) = u(x,t) - {}^{pc}/_2[u(x+h,t)-u(x-h,t)] + T,$$

where the truncation error, T, is $O(k^2+kh^2)$. As h and k tend to zero this gives the consistent difference method

$$U_m^{n+1} = U_m^n - {}^{kc}/_{2h}[U_{m+1}^n - U_{m-1}^n], \qquad (5.2.7)$$

which has local **discretization** error proportional to terms of $O(k+h^2)$. The domain of dependence for this method is shown in Figure 5.2.3 and so this method satisfies the CFL criterion provided the slope of the mesh is less than the slope of the characteristic, i.e. ${}^k/_h \leq {}^1/_c$.

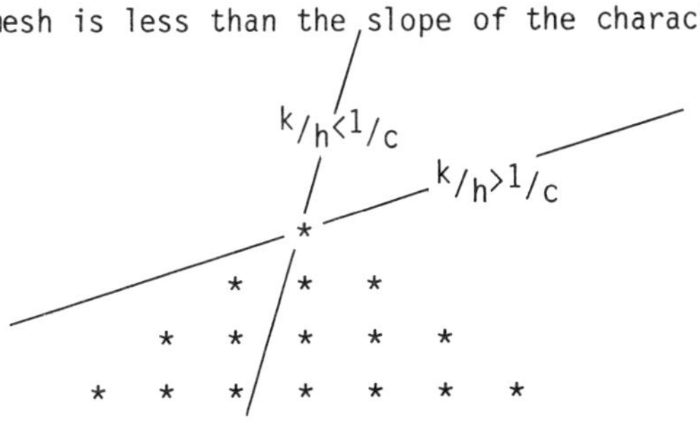

Figure 5.2.3 The Domain of Dependence for Equation (5.2.7).

The stability of the method (5.2.7) may be examined by considering solutions of the form $U_m^n = A\rho^n e^{im\phi}$. Substituting U_m^n into (5.2.7)

$$\rho = 1 - pc(e^{i\phi} - e^{-i\phi}) = 1 - i\ pc\ \sin\phi.$$

Therefore, $|\rho|>1$, except when $\phi=0$, hence $|U_m^n|$ will increase and we conclude that (5.2.7) is unstable despite satisfying the CFL criterion. This demonstrates that the CFL criterion, though necessary, is not sufficient to guarantee stability. However, if the differential equation, $u_t+cu_x=0$, has a constant coefficient, c, then $u_{tt} = (u_t)_t = (-cu_x)_t = -c(u_t)_x = -c(-cu_x)_x = c^2u_x$ and (5.2.6) becomes

$$u(x,t+k) = u(x,t) - kcu_x + {}^{(kc)^2}/_{2}u_{xx} + O(k^3).$$

Replacing u_x as before and u_{xx} by $[u(x+h,t)-2u(x,t)+u(x-h,t)]/_{h^2}$ gives the **Lax-Wendroff** method

$$U_m^{n+1}=U_m^n - {}^{kc}/_{2h}[U_{m+1}^n-U_{m-1}^n] + {}^{(kc)^2}/_{2h^2}[U_{m+1}^n-2U_m^n+U_{m-1}^n]. \tag{5.2.8}$$

The local discretization error of (5.2.8) is given by

$$E(h,k) = {}^1/_2ku_{tt} - {}^1/_2kc^2u_{xx} + O(k^2+h^2),$$

but as $u_t=-cu_x$ we have $u_{tt}=c^2u_{xx}$ and so $E=O(k^2+h^2)$. The Lax-Wendroff method uses the same points as the unstable method (5.2.7) and, therefore, satisfies the CFL condition provided ${}^{kc}/_h\leq 1$. However, unlike equation (5.2.7), the Lax-Wendroff method is stable provided $0<{}^{kc}/_h\leq 1$.

Exercise 5.2.4 Confirm that the Lax-Wendroff method (5.2.8) is consistent of order h^2+k^2 with the differential equation (5.2.1). Show that this method is stable provided $0<{}^{kc}/_h\leq 1$.

The Lax-Wendroff method is explicit and so its application is relatively straightforward provided we deal with any boundaries in a consistent way.

Example 5.2.3 Consider the solution of the differential equation $u_t+cu_x=0$, $c>0$, $0\leq x\leq 1$, $t\geq 0$, subject to the initial condition $u(x,0)=\psi(x)$, $0\leq x\leq 1$, and the boundary condition $u(0,t)=\phi(t)$, $t\geq 0$.

Solution. We apply the Lax-Wendroff method to the mesh shown in Figure 5.2.4.

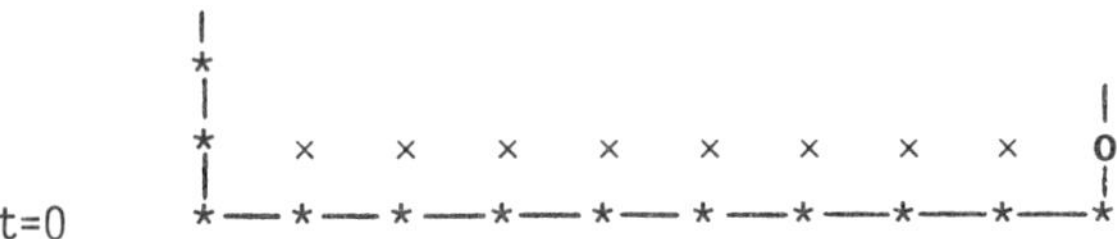

Figure 5.2.4 The Lax-Wendroff Method to Example 5.2.3

The initial and boundary conditions supply values at the point marked by "*" and the Lax-Wendroff scheme gives those marked "×" but the final value "**o**" cannot be found from either. This boundary value could be determined by using either the explicit first order method (5.2.3) or the first order implicit scheme (5.2.7) but this would detract from the overall second order accuracy of the Lax-Wendroff method (5.2.8). An alternative method is due to **Wendroff** and employs an average for u_t and u_x at suitable points. The differential equation (5.2.1) is replaced by

$$0.5([U_{m+1}^{n+1}-U_{m+1}^{n}]/k + [U_m^{n+1}-U_m^{n}]/k)+ 0.5c([U_{m+1}^{n+1}-U_m^{n+1}]/h + [U_{m+1}^{n}-U_m^{n}]/h)=0$$

which may be written

$$\alpha U_{m+1}^{n+1} + U_m^{n+1} = U_{m+1}^{n} + \alpha U_m^{n}, \qquad (5.2.9)$$

where $\alpha=(h+kc)/(h-kc)$. This method is implicit, by definition, but when used in conjunction with the boundary conditions or to determine the point marked o in connection with the Lax-Wendroff method it is in practice explicit. If used alone Wendroff's method satisfies the CFL criterion provided c>0 and is unconditionally stable.

Exercise 5.2.5 Show that Wendroff's method (5.2.9) is consistent of order (h^2+k^2). If ρ is the amplification factor associated with Wendroff's method show that $|\rho|=1$, for all possible meshes, and hence this method is unconditionally stable.

The amplification factor ρ associated with Wendroff's method has unit modulus and so this method is therefore always stable in theory. However, in practice any slight perturbation due to inexact arithmetic may shift ρ outside the unit circle which will produce an increasing solution and so Wendroff's method is said to be **weakly** stable.

Exercise 5.2.6 Use the matrix stability analysis to show that the combination of Lax-Wendroff and Wendroff's method used in Example 5.2.3 is stable if $0<{}^{kc}/_{h}\leq 1$.

Example 5.2.4 Compute a solution of the differential equation $u_t-2u_x=0$, $0\leq x\leq 1$, $t\geq 0$, subject to $u(x,0)=0$, $0\leq x\leq 1$, $u(0,t)=1$, $t\geq 0$. The solution of this problem is discontinuous across the characteristic $x=2t$; the following example considers how the Lax-Wendroff method copes with this problem.

Solution. We apply the Lax-Wendroff method with the following mesh sizes:-

$$\text{(a) } {}^{kc}/_{h}=1, \quad \text{(b) } {}^{kc}/_{h}=0.5<1.$$

Table 5.2.1. Computed Solution for Example 5.2.4.
$u_t-2u_x=0$, $0\leq x\leq 1$, $t\geq 0$, $u(x,0)=0$, $u(0,t)=1$.

	u(x,0.05)			u(x,0.1)		
x	(a)	(b)	(c)	(a)	(b)	(c)
0.0	0.0000	0.0000	0.0000	0.0000	0.0000	0.0000
0.1	1.0000	0.6250	0.5000	0.0000	0.1504	0.1250
0.2	1.0000	1.0000	1.0000	1.0000	0.6484	0.5000
0.3	1.0000	1.0000	1.0000	1.0000	0.9473	0.8750
0.4	1.0000	1.0000	1.0000	1.0000	1.0000	1.0000
0.5	1.0000	1.0000	1.0000	1.0000	1.0000	1.0000
0.6	1.0000	1.0000	1.0000	1.0000	1.0000	1.0000
0.7	1.0000	1.0000	1.0000	1.0000	1.0000	1.0000
0.8	1.0000	1.0000	1.0000	1.0000	1.0000	1.0000
0.9	1.0000	1.0000	1.0000	1.0000	1.0000	1.0000
1.0	1.0000	1.0000	1.0000	1.0000	1.0000	1.0000

Table 5.2.2. Computed Solution for Example 5.2.4.
$u_t-2u_x=0$, $0<x<1$, $t>0$, $u(x,0)=0$, $u(0,t)=1$.

x	u(x,0.15) (a)	(b)	(c)	u(x,0.20) (a)	(b)	(c)
0.0	0.0000	0.0000	0.0000	0.0000	0.0000	0.0000
0.1	0.0000	-0.0292	0.0313	0.0000	-0.0472	0.0078
0.2	0.0000	0.2266	0.1875	0.0000	-0.0192	0.0625
0.3	1.0000	0.6581	0.5000	0.0000	0.2746	0.2265
0.4	1.0000	0.9209	0.8125	1.0000	0.6632	0.5000
0.5	1.0000	0.9926	0.9688	1.0000	0.9039	0.7734
0.6	1.0000	1.0000	1.0000	1.0000	0.9847	0.9375
0.7	1.0000	1.0000	1.0000	1.0000	0.9990	0.9922
0.8	1.0000	1.0000	1.0000	1.0000	1.0000	1.0000
0.9	1.0000	1.0000	1.0000	1.0000	1.0000	1.0000
1.0	1.0000	1.0000	1.0000	1.0000	1.0000	1.0000

From Tables 5.2.1 and 5.2.2 it is clear that when the slope of the mesh coincides with the slope of the characteristic then the exact solution is determined; an examination of the local discretization error of the Lax-Wendroff method gives the reason for this phenomenon. If the step size is chosen as in (b) then any discontinuity will be "blurred" out as shown in this example. This is also the case if any other method is used. The column (c) is determined using the simple explicit method (5.2.3) with $kc/h=0.5$. The degree to which the sharp wave front is spread out depends upon the form of the local discretization error. A full discussion is given by Richtmyer and Morton (1967). It is clear that the Lax-Wendroff method, which is of second order in both k and h, deals with the discontinuity better that the simple explicit method.

The extension of difference methods for solving $u_t+cu_x=0$ to those which involve more than two levels of the mesh is straightforward but the analysis of such methods and the way in which they are implemented is not.

Exercise 5.2.7 The **Leapfrog Method** for solving (5.2.1) is obtained by replacing u_x and u_t by central difference approximations to give

$$[U_m^{n+1} - U_m^{n-1}]/2k + c[U_{m+1}^n - U_{m-1}^n]/2h = 0.$$

Show that this method (i) has local discretization error $O(h^2,k^2)$,
(ii) satisfies the Courant-Friedrichs-Lewy condition if $0<kc/h<1$,
(iii) is weakly stable,
(iv) requires a suitable starting procedure as it involves 3 levels of the mesh,
(v) the solution on a mesh $\Gamma_{h,k}$ is made up of two solutions on disjoint meshes of the form $\Gamma_{2h,2k}$.

5.2.1 Variable Coefficients

The extension of difference methods to deal with equations of the form

$$u_t + cu_x = 0, \; (x,t) \text{ in } \Gamma = \{ x,t > 0 \},$$

where c is now a variable coefficient, is accomplished by replacing the constant c in the previous methods with its value at the point at which the differential equation is approximated.

Example 5.2.5 (a) The simple explicit method (5.2.3) becomes

$$[U_m^{n+1}-U_m^n]/k + c_m^n[U_m^n-U_{m-1}^n]/h = 0,$$

where $c_m^n = c(mh,nk,U_m^n)$, i.e. since this method is based on an approximation to $u_t+c(x,t,u)u_x=0$ at $x=mh$, $t=nk$, the coefficient c is replaced by c_m^n. This method is $O(h+k)$ and locally stable provided $0<kc_m^n/h<1$.

(b) The simple implicit method (5.2.4) is based an an approximation for $u_t+cu_x=0$ at the point $x=mh,t=(n+1)k$ and so c is replaced by c_m^{n+1}. This results in a method which has discretization error $O(h+k)$ and is unconditionally stable provided $c>0$.

(c) Wendroff's method (5.2.9) is based upon an averaging process relating the four points (mh,nk), (mh,(n+1)k), ((m+1)h,nk) and ((m+1)h,(n+1)k) and so c is replaced by $c_{m+\frac{1}{2}}{}^{n+\frac{1}{2}}$. If c is independent of U then $c((m+1/2)h,(n+1/2)k)$ is used, if not then it is necessary to interpolate U at the four mesh points around $x=(m+1/2)h$, $t=(n+1/2)k$. The resulting method is still second order and locally stable provided $c>0$.

Exercise 5.2.8 Show that (a) - (c) given in Example 5.2.5 are of the same order as achieved in the constant coefficient case.

The extension of the Lax-Wendroff method (5.2.8) is a little more complicated since it depends upon replacing higher order derivatives and will be considered in Section 5.2.3.

5.2.2 Inhomogeneous First Order Hyperbolic Equations

Let us now consider the solution of the inhomogeneous problem

$$u_t + cu_x = g, \quad x,t\geqslant 0, \qquad (5.2.10)$$

subject to a suitable initial condition at t=0 and boundary conditions at x=0. Each method discussed can be extended to such a problem as follows.

Example 5.2.6 (a) The simple explicit method (5.2.3) becomes

$$[U_m{}^{n+1}-U_m{}^n]/k + c[U_m{}^n-U_{m-1}{}^n]/h = g_m{}^n,$$

where $g_m{}^n=g(mh,nk)$.

(b) Wendroff's method (5.2.9) is given by replacing the right-hand side by g evaluated at $x=(m+1/2)h$, $t=(n+1/2)k$.

Exercise 5.2.9 At which point will it be necessary to evaluate g for the simple implicit method (5.2.4)?

Exercise 5.2.10 Examine the consistency and stability of the methods suggested in Example 5.2.6 and Exercise 5.2.8.

5.2.3 Non-linear First Order Hyperbolic Equations

Many physical problems can be written in the conservative form

$$u_t + \partial F(u)/\partial x = 0. \qquad (5.2.11)$$

This equation can be re-written as

$$u_t + \partial F/\partial u \; u_x = 0$$

which may be approximated as in Section 5.2.2, but it is more convenient to replace (5.2.11) directly. Only the Lax-Wendroff method will be considered but a similar approach can be applied to each of the other explicit or implicit methods.

The basis of the Lax-Wendroff method is to expand the solution at x,t+k in a power series about x,t, i.e.

$$\begin{aligned} u(x,t+k) &= u(x,t) + ku_t + k^2/2u_{tt} + \ldots\ldots. \\ &= u(x,t) - k\,\partial F/\partial x - 1/2k^2\,\partial/\partial t[\partial F/\partial x]. \end{aligned} \qquad (5.2.12)$$

The next step is to replace $\partial/\partial t[\partial F/\partial x]$ as follows

$$\partial/\partial t[\partial F/\partial x] = \partial/\partial x[\partial F/\partial t] = \partial/\partial x[-\partial F/\partial u\;\partial u/\partial x] = -\partial/\partial x[\,A\;\partial u/\partial x],$$

where $A = \partial F/\partial u$, and then (5.2.12) becomes

$$\begin{aligned} U_m^{n+1} &= U_m^n - kc/2h[F_{m+1}^n - F_{m-1}^n] \\ &\quad + k^2/2h^2[A_{m+\frac{1}{2}}^n(F_{m+1}^n - F_m^n) - A_{m-\frac{1}{2}}^n(F_m^n - F_{m-1}^n)], \end{aligned} \qquad (5.2.13)$$

where A_m^n is A evaluated at x=mh t=nk. This is an explicit method which is of second order in h and k and is locally stable provided the CFL criterion is satisfied. The values $A_{m\pm\frac{1}{2}}$ are frequently replaced by $[A_{m+1}^n + A_m^n]/2$ and $[A_m^n + A_{m-1}^n]/2$, respectively, to avoid additional interpolation.

Exercise 5.2.11 Show that the difference method (5.2.13) has discretization error $O(h^2+k^2)$ and is locally stable provided the CFL criterion is satisfied.

5.2.4 Systems of Hyperbolic Equations

All numerical methods developed extend to consider the solution of the system of N first order hyperbolic equations

$$\mathbf{U}_t + \mathbf{A}\mathbf{U}_x = 0, \qquad (5.2.14)$$

subject to suitable initial and boundary conditions, where $\mathbf{U}=[u_1(x,t),\ldots,u_N(x,t)]^T$ and A is an N dimensional square matrix. The slopes of the characteristics are given by the eigenvalues of A. For the sake of simplicity we shall assume that all the eigenvalues of A are positive which means that all the characteristics traverse the x-t plane in the same sense.

Example 5.2.7 The simple explicit method (5.2.3) becomes

$$[\mathbf{U}_m{}^{n+1} - \mathbf{U}_m{}^n]/k + \mathbf{A}[\mathbf{U}_m{}^n - \mathbf{U}_{m-1}{}^n]/h = 0, \qquad (5.2.15)$$

where $\mathbf{U}_m{}^n$ is an approximation for $\mathbf{U}(mh,nk)$, which is stable provided the step size k is chosen such that $0<k\lambda_i/h\leq 1$, where λ_i, $i=1,\ldots,N$, are the eigenvalues of A. If A has non-positive eigenvalues then it is necessary to partition the equation (5.2.14) in such a way that (5.2.15) is applied to those components of U which correspond to the positive eigenvalues whilst the equivalent of (5.2.2) is applied to the rest.

Exercise 5.2.12 Consider the stability of (5.2.15) by assuming that A is similar to a diagonal matrix, i.e. there exists a non-singular matrix P such that $\mathbf{A}=\mathbf{P}\mathbf{D}\mathbf{P}^{-1}$, where the entries of D are the eigenvalues of A.

Example 5.2.8 The generalisation of the Lax-Wendroff method is given by (5.2.13) in which $U_m{}^n$ is replaced by $\mathbf{U}_m{}^n$ and $F_m{}^n$ by $[\mathbf{AU}]_m{}^n$.

Exercise 5.2.13 Confirm that the Lax-Wendroff method as applied in Example 5.2.8 is consistent and locally stable if $0<{}^{k\rho(\mathbf{A})}/_h\leq 1$, where $\rho(\mathbf{A})$ is the spectral radius of A.

5.3 The Method of Characteristics

As the characteristics provide the natural directions along which to integrate a hyperbolic equation it would seem sensible to use them whenever possible. From Section 5.1 the slopes of the characteristics associated with the equation $u_t+cu_x=f(x,t,u)$ are given by those directions along which the equations

$$\begin{aligned} u_t + cu_x &= f, \\ u_t dt + u_x dx &= D(u) \end{aligned} \qquad (5.3.1)$$

are singular. From equation (5.1.7) this gives dx-cdt=0 and this equation may sometimes be integrated to determine an explicit form for the characteristics. More frequently it is not possible to integrate this equation exactly and so it is necessary to resort to a numerical method; this is particularly the case when c depends on the solution u, i.e.

$$^{dx}/_{dt} = c(x,u,t). \qquad (5.3.2)$$

However, equation (5.3.1) will have a unique solution along the characteristics if

$$\text{Det}\begin{bmatrix} f & 1 \\ D(u) & dt \end{bmatrix} = 0, \qquad (5.3.3)$$

i.e. fdt-D(u)=0 or $^{D(u)}/_{dt}=f$. Therefore, beginning at each initial point of the mesh, equations (5.3.2) and (5.3.3) are integrated to produce a difference solution.

Example 5.3.1 Compute a difference solution for the equation $u_t+uu_x=2x$ subject to $u(x,0)=\exp(-x)$, 0 x 1.

Solution. The interval (0,1) is divided up into a uniform mesh with spacing 0.1. Equations (5.3.2) and (5.3.3) become

$$^{dx}/_{dt} = u \text{ and } ^{D(u)}/_{dt} = 2x.$$

Let us suppose that at some time, t_{old}, the corresponding values for x and u are x_{old} and u_{old} then using Euler's method to integrate the above equations over a time interval k give updated values x_{new} and u_{new} as follows:-

$$x_{new} = x_{old} + ku_{old},$$
$$u_{new} = u_{old} + 2kx_{old}.$$

For example, beginning at t=0, with $x_{old}=0$ where $u_{old}=\exp(-x_{old})=1$ produces the values $x_{new}=0.1$ and $u_{new}=1.0$. Repeating this for x=0.1,0.2,...,1.0 gives the results shown in Table 5.3.1 below.

Table 5.3.1 The solution at time t=0.1 for Example 5.3.1.

x_{old}	x_{new}	u_{new}	x_{exact}	$u(x_{exact},0.1)$
0.0	0.100000	1.000000	0.100334	1.010017
0.1	0.190484	0.924837	0.191787	0.933968
0.2	0.281873	0.858730	0.284150	0.867065
0.3	0.374082	0.800818	0.377334	0.808439
0.4	0.467032	0.750320	0.471262	0.757301
0.5	0.560653	0.706531	0.565864	0.712940
0.6	0.654881	0.668812	0.661074	0.674709
0.7	0.749659	0.636585	0.756836	0.642027
0.8	0.844933	0.609329	0.853096	0.614364
0.9	0.940657	0.586570	0.949808	0.591243
1.0	1.036788	0.567880	1.046927	0.572232

Figure 5.3.1 The difference mesh determined for Example 5.3.1.

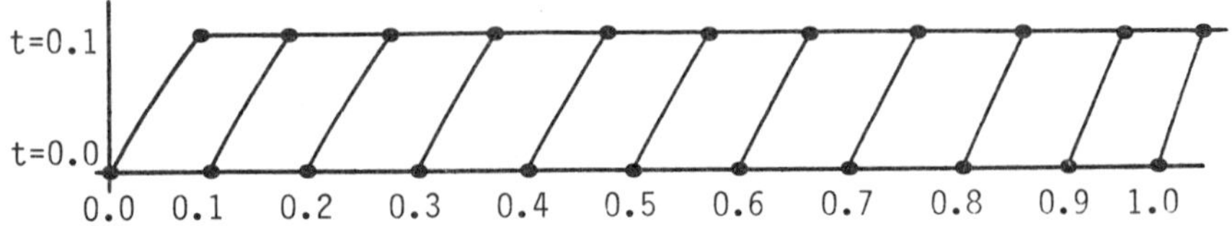

Clearly, if a more accurate scheme were employed to integrate equation (5.3.2) and (5.3.3) the difference solution would also be improved. Notice that the mesh on which the solution is computed will migrate away from the boundary x=0. However, if u is specified at x=0 then it is possible to follow the characteristics which meet this line by integrating the characteristic equations in the form

$$dt/dt = 1/u, \text{ and } D(u)/dx = D(u)/dt \; dt/dx = 2x/u,$$

subject to u(0) and t(0) being given at x=0. Notice, also, that it is not necessary to compute the solution over the whole region $x \geq 0$, $t \geq 0$, but that it is possible to follow individual characteristics.

Exercise 5.3.1 Show that using the Classical Runge-Kutta method produces the exact solution given in Table 5.3.1.

Due to the explicit way in which the characteristics of the problem have been used this method has many advantages, but suffers from two major drawbacks. Firstly, a lot more effort is required to compute the solution and keep track of the points at which the approximate solution is determined, and secondly the mesh is constantly changing shape. This second difficulty can be overcome either by interpolating the solution at each level to conform to a fixed mesh or else by fixing the mesh at the $n+1^{th}$ level of the mesh and drawing the characteristics back to the nth level at which point this solution is interpolated. (See Figure 5.3.2) This is usually referred to as a Hartree hybrid method, further details of which are given by Ames (1977).

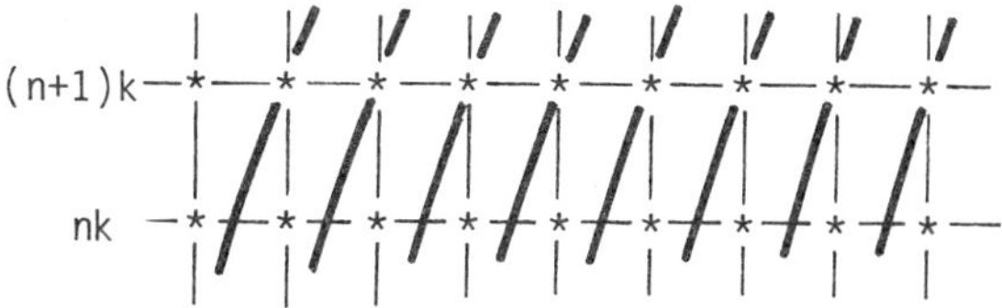

Figure 5.3.2 The characteristics from the n+1th level of $\Gamma_{h,k}$.

Exercise 5.3.2 Compute a solution for the problem given in Example 5.3.1 using a fixed mesh $\Gamma_{h,k}$.

5.4 Second Order Hyperbolic Equations

The wave equation

$$u_{tt} = c^2 u_{xx},\ 0 \leqslant x \leqslant 1,\ t \geqslant 0, \tag{5.4.1}$$

subject to the initial conditions $u(x,0)=\phi(x)$ and $u_t(x,0)=\psi(x)$, and boundary conditions at x=0 and x=1, can be used as a non-dimensional model for many physical situations. Equation (5.4.1) is the simplest example of a second order linear hyperbolic equation and is in the canonical form as described in Appendix 2. If c is a constant, then from Appendix 2, the characteristic lines are given by $x-ct=\xi$ and $x+ct=\eta$ where ξ and ν are constants for specific characteristics. Changing to the characteristic variables ξ and η equation (5.4.1) can be written in the alternative canonical form

$$\partial^2 u/\partial\xi\partial\eta = 0, \tag{5.4.2}$$

the solution of which is given by $u(\xi,\eta)=F(\xi)+G(\eta)$, where F and G are arbitrary functions of a single variable. Therefore, the general solution, usually called D'Alembert's solution, of (5.4.1) is given by

$$u(x,t) = F(x-ct) + G(x+ct). \tag{5.4.3}$$

Consider the solution of (5.4.1), subject to the initial values $u(x,0)=\phi(x)$ and $u_t(x,0)=\psi(x)$, $0 \leqslant x \leqslant 1$. The characteristics through the point A intersect the initial line, t=0, at the points C and D and it is possible to show that the solution at this point is given by

$$u(x,t) = {}^1/_2\, [\phi(x-ct)+\phi(x+ct)] - \frac{1}{2c}\int_C^D \psi(s)ds, \qquad (5.4.4)$$

i.e. the solution at A is uniquely determined by the values of ϕ and ψ between the points C and D. However, at the point B one of the characteristics meets the line t=0 outside the interval (0,1) and though it is possible to extend the initial conditions or reflect the characteristic back from the boundary at x=1 the determination of the solution at this point is complicated. If c is not a constant then the extension of this approach is not always possible and so we shall consider numerical methods as an alternative.

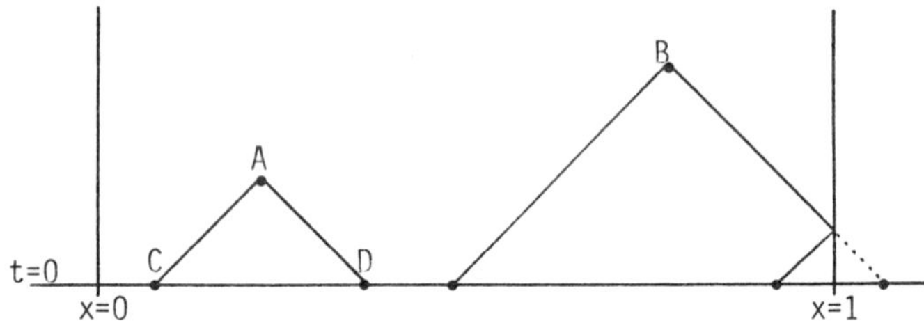

Figure 5.4.1 The characteristics passing through points A **and** B.

5.4.1 Finite Difference Methods

In order to construct a solution for equation (5.4.1) the region $\Gamma = \{(x,t) : 0 \leq x \leq 1,\ t \geq 0\}$ is covered by a mesh $\Gamma_{h,k}$ where Mh=1. At each point of $\Gamma_{h,k}$ we determine a value U_m^n which is an approximation for u(mh,nk). The simplest difference approximation for $u_{tt}=c^2u_{xx}$ is given by

$$\frac{\delta_t^2 U_m^n}{k^2} = c^2\,\frac{\delta_x^2 U_m^n}{h^2}, \qquad (5.4.5)(i)$$

or

$$\frac{(U_m^{n+1}-2U_m^n+U_m^{n-1})}{k^2} = c^2\frac{(U_{m+1}^n-2U_m^n+U_{m-1}^n)}{h^2}. \qquad (5.4.5)(ii)$$

This method has local discretization error $O(k^2+h^2)$ and is therefore consistent with the differential equation (5.4.1). The domain of dependence of a point x=mh, t=nk of the mesh $\Gamma_{h,k}$ is shown in Figure 5.4.2. Therefore, the CFL criterion is satisfied provided $0<k^2c^2/h^2\leq 1$ which is also sufficient to guarantee that the difference method (5.4.5) is stable. However, as this method involves 3 levels of the mesh, it is necessary to use a special procedure in order to start the recurrence. The initial conditions associated with the equation $u_{tt}=c^2u_{xx}$ are given by $u(x,0)=\phi(x)$ and $\partial u/\partial t(x,0)=\psi(x)$, therefore, from the Taylor series expansion

$$u(x,k)=u(x,0) + k\partial u/\partial t + k^2/2\ \partial^2u/\partial t^2 + \ldots.$$

we obtain

$$U_m^1 = U_m^0 + k\psi(mh) + k^2/2\ c^2\phi''(mh)$$

to $O(k^2)$. The difference method (5.4.5) may then be used to determine U_m^n, m=1,..,M-1, for n>1.

Figure 5.4.2 The Domain of Dependence for equation (5.4.5).

If derivative boundary conditions are specified at x=0 and/or x=1 then we can approximate them by extending the mesh and introducing "fictitious" mesh values. By approximating the boundary condition and using a suitable approximation for the differential equation the additional fictitious values can be eliminated. (See Example 4.4.1 regarding a similar problem with parabolic partial differential equations.)

The stability restriction imposed upon the explicit method (5.4.5) may be overcome by using a suitable implicit method. For example,

$$\delta_t^2 U_m{}^n/k^2 = 0.25[\delta_x^2 U_m{}^{n+1} + 2\delta_x^2 U_m{}^n + \delta_x^2 U_m{}^{n-1}]/h^2 \qquad (5.4.6)$$

uses a weighted average for u_{xx} at $t=(n-1)k,nk$, and $(n+1)k$. This gives a consistent unconditionally stable method. The difference method (5.4.6) requires the solution of a tridiagonal system of equations at each step, but these are diagonally dominant provided the problem is properly posed. Higher order schemes have been proposed such as those described in Mitchell (1969), Jain (1978) and Smith (1978) but these are rarely used as most require complicated modification at the boundaries. The generalisation of both explicit and implicit methods to more general second order equations is carried out by replacing any derivative by suitable difference approximations, to ensure a consistent method, and then checking for local stability.

5.4.2 The Method of Characteristics applied to the Wave Equation

We suppose that the solution of (5.4.1) is known at some point x,t and investigate how the solution at a neighbouring point x+h, t+k is related to it. The directional derivatives of u_t and u_x in the direction of the line segment joining these two points is given by

$$D(u_t) = u_{tx}dx + u_{tt}dt, \qquad (5.4.7)(i)$$

$$D(u_x) = u_{xx}dx + u_{xt}dt, \qquad (5.4.7)(ii)$$

which combined with the differential equation $u_{tt}=c^2u_{xx}$ can be collectively written as

$$\begin{bmatrix} 1 & 0 & -c^2 \\ 0 & dt & dx \\ dt & dx & 0 \end{bmatrix} \begin{bmatrix} u_{tt} \\ u_{xt} \\ u_{xx} \end{bmatrix} = \begin{bmatrix} 0 \\ D(u_x) \\ D(u_t) \end{bmatrix}.$$

The characteristic directions are given by $dx^2-c^2dt^2=0$ which, if c is a constant, may be integrated to give the lines $x-ct=\xi$ and $x+ct=\eta$. Along these lines there can be no solution unless we also have that

$$\text{Det}\begin{bmatrix} 1 & 0 & 0 \\ 0 & dt & D(u_x) \\ dt & dx & D(u_t) \end{bmatrix} = 0$$

or

$$D(u_t)\ dt \ - D(u_x)\ dx = 0. \qquad (5.4.8)$$

This equation, together with $dx=\pm cdt$, can be integrated numerically to provide an approximate solution as follows.

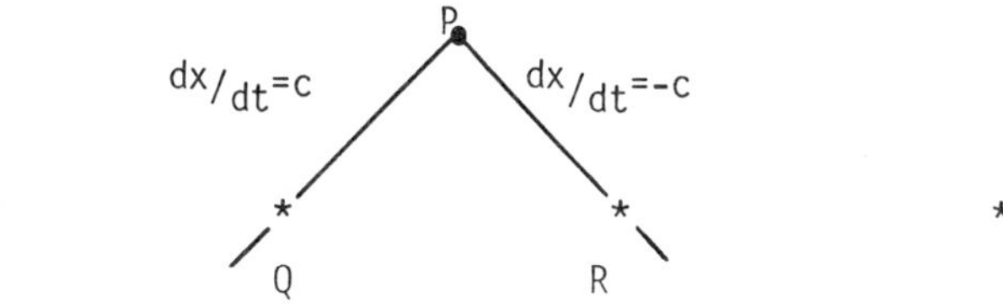

Figure 5.4.3 The Application of Method of Characteristics to the Wave Equation $u_{tt}=c^2u_{xx}$, $0\leqslant x\leqslant 1$, $t\geqslant 0$.

Let us suppose that the solution is known at a set of points, not necessarily uniformly distributed, at t=nk as shown in Figure 5.4.3. From the points Q and R we integrate ${}^{dx}/_{dt}=c$ and ${}^{dx}/_{dt}=-c$ respectively. (If c is constant then the characteristics are determined analytically to give $x\pm ct=$constant.). If the characteristics from R and Q meet at P then integrating (5.4.8) gives the following:

From Q: $(dx=+cdt)$ $\quad D(u_t) - c\ D(u_x) = 0$, which to first order gives

$$(u_t|_P - u_t|_Q) - c(u_x|_P - u_x|_Q) = 0, \qquad (5.4.9)$$

where $u_t|_P$ and $u_x|_P$ denote the values of $\partial u/\partial t$ and $\partial u/\partial x$ at the point P.

From R: (dx=-cdt) $\quad D(u_t) + c\, D(u_x) = 0$

or
$$(u_t|_P - u_t|_R) + c(u_x|_P - u_x|_R) = 0. \tag{5.4.10}$$

which may be solved for $u_t|_P$ and $u_x|_P$. Clearly a more accurate approximation for (5.4.8) would give improved results. Having found u_t and u_x we integrate to find u at R.

The use of equations (5.4.9) and (5.4.10) to solve the simple wave equation suffers from a non-uniform, mobile mesh but this can be overcome by interpolating the computed solution at the $n+1^{th}$ level of the mesh to give a uniform spacing or else by selecting the mesh at the $n+1^{th}$ level to be uniform and then tracing the characteristics from each point x=mh, t=(n+1)k back to the nth level of the mesh as was described in Section 5.3.

5.4.3 The Wave Equation as a System of First Order Equations

The wave equation $u_{tt}=c^2u_{xx}$ may be written as a system of first order equations given by

$$\begin{aligned} p_t - cq_x &= 0, \\ q_t - cp_x &= 0, \end{aligned} \tag{5.4.11}$$

where $p=u_t$ and $q=u_x$. These equations may then be written in the form

$$\partial U/\partial t + \partial/\partial x(\mathbf{F}(\mathbf{U})) = 0, \tag{5.4.12}$$

where $U=[p,q]^T$ to which methods such as the Lax-Wendroff scheme could be applied. However, due to the importance of the wave equation special schemes have been developed for equations (5.4.11).

To demonstrate that the most obvious schemes are not always suitable consider the following approximation for the differential equations (5.4.11).

$$(p_m^{n+1} - p_m^{n})/k = c(q_{m+1}^{n} - q_{m-1}^{n})/2h$$
$$(q_m^{n+1} - q_m^{n})/k = c(p_{m+1}^{n} - p_{m-1}^{n})/2h \tag{5.4.13}$$

The difference equations (5.4.13) are consistent, of order 1, with the differential equations (5.4.11). To examine the stability of this scheme let us consider solutions of the form $p_m^n = A\rho^n \exp(im\phi)$ and $q_m^n = B\rho^n \exp(im\phi)$. Substituting these expressions into the above difference equations gives

$$2A(\rho-1) = {}^{kc}/_{h}\, B(e^{i\phi} - e^{-i\phi})$$
$$2B(\rho-1) = {}^{kc}/_{h}\, A(e^{i\phi} - e^{-i\phi})$$

and eliminating ${}^{A}/_{B}$ produces $\rho = 1 \pm i\,{}^{kc}/_{h} \sin\phi$. Therefore, $|\rho| > 1$ and we conclude that $|p_m^n|$ and $|q_m^n|$ are unbounded. If c is not a constant then the solution of the differential equation may be increasing but as $\rho > 1 + O(k)$ the difference solution is still unstable.

Exercise 5.4.1 Show that the explicit difference approximations

(a) $(p_m^{n+1} - (p_{m+1}^{n} + p_{m-1}^{n})/2)/k = c(q_{m+1}^{n} - q_{m-1}^{n})/2h,$

$(q_m^{n+1} - (q_{m+1}^{n} + q_{m-1}^{n})/2)/k = c(p_{m+1}^{n} - p_{m-1}^{n})/2h,$

(b) $(p_m^{n+1} - p_m^{n})/k = c(q_{m+\frac{1}{2}}^{n} - q_{m-\frac{1}{2}}^{n})/h,$

$(q_{m-\frac{1}{2}}^{n+1} - q_{m-\frac{1}{2}}^{n})/k = c(p_m^{n+1} - p_{m-1}^{n+1})/h,$

are both stable provided $0 < {}^{kc}/_{h} \leq 1$. Determine which method is the most accurate.

Exercise 5.4.2 Show that the implicit difference method, for solving (5.4.11), given by

$$(p_m^{n+1} - p_m^n)/k = c(q_{m+\frac{1}{2}}^n - q_{m-\frac{1}{2}}^n + q_{m+\frac{1}{2}}^{n+1} - q_{m-\frac{1}{2}}^{n+1})/2h,$$

$$(q_{m-\frac{1}{2}}^{n+1} - q_{m-\frac{1}{2}}^n)/k = c(p_m^{n+1} - p_{m-1}^{n+1} + p_m^n - p_{m-1}^n)/2h$$

is unconditionally stable and determine its order.

5.5 Concluding Remarks Regarding Hyperbolic Equations

The extension of the methods discussed in this chapter to equations in more than one spatial dimension, i.e. differential equations of the form

$$u_{tt} = c^2(u_{xx} + u_{yy}), \qquad (x,y) \text{ in } \Gamma \subset R^2,\ t>0,$$

subject to the initial conditions $u(x,y,0)=f(x,y)$ and $\partial u/\partial t(x,y,0)=g(x,y)$, for (x,y) in Γ, and boundary conditions specified on $\partial\Gamma\times(0,T)$, where $\partial\Gamma$ is the boundary of Γ, may be carried out in the obvious way. The region $\Gamma\times(0,T)$ is covered with a finite difference mesh $\Gamma_{h,k}=\{\ x=mh,\ y=\ell h,\ t=nk\ :\ (x,y) \text{ in } \Gamma,\ 0\leq t\leq T\ \}$ at each point of which $U_{m\ell}^n \simeq u(mh,\ell h,nt)$. The differential equation is then approximated by

$$\frac{\delta_t^2 U_{m\ell}^n}{k^2} = c^2 \ \frac{(\delta_x^2 U_{m\ell}^n + \delta_y^2 U_{m\ell}^n)}{h^2} \qquad (5.5.1)$$

subject to $U_{m\ell}^n$ being specified on the boundary $\partial\Gamma\times(0,T)$. The local discretization error associated with this approximation can be found by expanding in a Taylor series about $x=mh$, $y=\ell h$, $t=nk$ and stability analysed using the von Neumann method. A generalisation to implicit and ADI methods can be carried out as in Chapter 4.

Exercise 5.5.1. Determine the principal term in the local discretization error associated with the simple explicit method (5.5.1). For which values of k and h is the difference function $U_{m\ell}{}^{n}=\rho^{n}\exp(im\phi)\exp(i\ell\phi)$ a bounded solution of (5.5.1).

The propagation of discontinuities is a fundamental consequence of the existence of characteristics associated with hyperbolic partial differential equations. For example, the differential equations describing the flow of gases are based on various conservation laws, (mass, momentum and energy), and such flows are often exemplified by internal discontinuities, i.e. shocks, across which internal boundary conditions are required. In physical terms a shock is a very narrow region across which there are very rapid changes in variables which describe the pressure, velocity, density, temperature and entropy. Such rapid changes can be dealt with by considering the the region in which they take place as a surface. Across this surface additional physical conditions need to be specified. Ames (1977) gives examples for which the so called Rankine-Hugoniot jump conditions are applied using one sided derivatives on either side of the interface. If the position of the interface is known then it is possible to adjust the mesh so that the interface falls at a mesh point and then use the basic difference schemes which have been described taking into account any discontinuity at the interface node. However, this approach must be used with care since large changes in the solution may require widely differing mesh spacings on either side of the shock front. If possible the mesh size should be selected so that $^{h}/_{k}$=interface velocity but this can lead to instability. Von Neumann and Richtmyer (1950) developed the so called pseudo-viscosity method which attempts to simulate shock waves by introducing higher order terms into hyperbolic equations and used it with considerable success. (For an example see Richtmyer and Morton (1967)). As an alternative Lax (1954) developed a technique which does not introduce an explicit pseudo-viscosity but has the same effect by using suitable difference approximations. For example, instead of replacing u_t by $(U_m^{n+1}-U_m^n)/k$ it is approximated by

$$(U_m^{n+1} - 1/2\,[U_{m+1}^n + U_{m-1}^n])/k,$$

which may be written as

$$(U_m^{n+1} - U_m^n)/k - 1/2(U_{m+1}^n-2U_m^n+U_{m-1}^n)/k.$$

Expanding these terms in Taylor series about x=mh, t=nk gives $u_t+ku_{tt}+....+h^2/2k\,u_{xx}+...$ where the final term is the pseudo-viscosity introduced by von Neumann and Richtmyer. Examples of this approach are given in Exercise 5.4.1 (a).

CHAPTER 6

Elliptic Partial Differential Equations

6.1 Introduction

As a simple example of an elliptic equation recall that the Heat Equation in 2 spatial dimensions is given by

$$u_t = \sigma(u_{xx} + u_{yy}), \quad (x,y) \text{ in } \Gamma \subset R^2, \; t>0,$$

which if a steady state solution exists, i.e. $\partial u/\partial t=0$, reduces to

$$u_{xx} + u_{yy} = 0, \quad (x,y) \text{ in } \Gamma.$$

The latter is an elliptic partial differential equation and is known as Laplace's equation. It arises frequently in mathematical modelling associated with the steady flow of heat or electricity, with the irrotational flow of an incompressible fluid and with potential problems in electricity and magnetism. Elliptic partial differential equations are always of boundary value type. (See Appendix 2 for a simple classification of both partial differential equations and boundary value problems.) In order for a solution to be determined over the bounded region Γ it is necessary that either u, the normal derivative of u or a combination of both be given at each point of the boundary $\partial\Gamma$. If u is specified at all points of the boundary the determination of the solution is called a Dirichlet problem and it is possible to show, using Green's Theorem, that there exists a unique solution which depends continuously on all points of $\partial\Gamma$. (See Garabedian (1964), Weinberger (1965), Gahkov (1966).) If the outward normal derivative is specified on $\partial\Gamma$, a Neumann problem, then the solution is only unique to within a constant. The final type of problem, the solution of the differential equation subject to $\partial u/\partial n + a(x,y)u = g(x,y)$, on $\partial\Gamma$, is called a Robin or third boundary value problem and has a unique solution only if $a>0$.

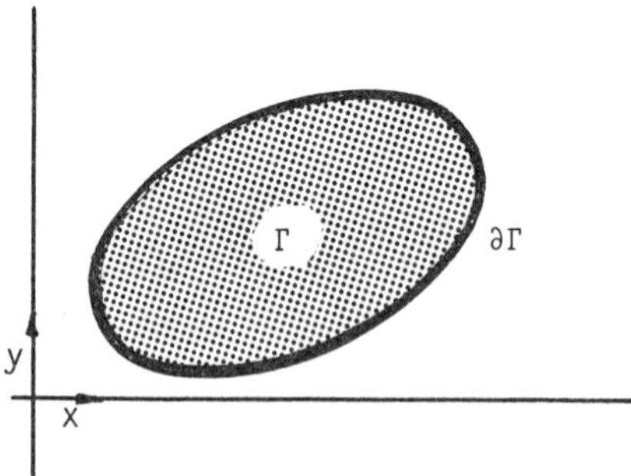

Example 6.1.1 Consider Laplace's equation for which a solution is required over the square $0 \leqslant x,y \leqslant 1$ subject to $u=0$ on the boundary, except for the line segment $0 \leqslant x \leqslant 1$, $y=0$ where $u(x,0)=\sin^2 \pi x$. It is possible to show that the solution of this problem is given by

$$u(x,y) = -8/\pi \sum_{j=0}^{\infty} \frac{\sinh(2j-1)\pi(1-y)}{(4j^2-1)(2j-3)\sinh(2j-1)\pi} \sin(2j-1)\pi x.$$

where the series converges uniformly for $0 \leqslant x,y \leqslant 1$. (See Weinberger (1965).). Given an elliptic equation which does not have constant coefficients, is inhomogeneous, is non-linear or is defined on a non-regular region the determination of an exact analytical solution is not always possible. It is frequently necessary to resort to numerical methods.

6.2 Simple Numerical Methods for Laplace's Equation on 0<x,y<1

We begin by considering the solution of Laplace's equation

$$u_{xx} + u_{yy} = 0, \quad (x,y) \text{ in } \Gamma = \{(x,y) : 0 \leq x,y \leq 1\}, \qquad (6.2.1)$$

subject to the boundary conditions

$$u(x,y) = \begin{cases} f_1(x), & y=0,\ 0 \leq x \leq 1, \\ f_2(x), & y=1,\ 0 \leq x \leq 1, \\ g_1(y), & x=0,\ 0 \leq y \leq 1, \\ g_2(y), & x=1,\ 0 \leq y \leq 1. \end{cases}$$

The region Γ is covered by a mesh, Γ_h, which has uniform spacing h. Figure 6.2.1 illustrates such a mesh with $h={}^1/_3$. It is possible to use different spacings in the x and y directions but this will not be considered here.

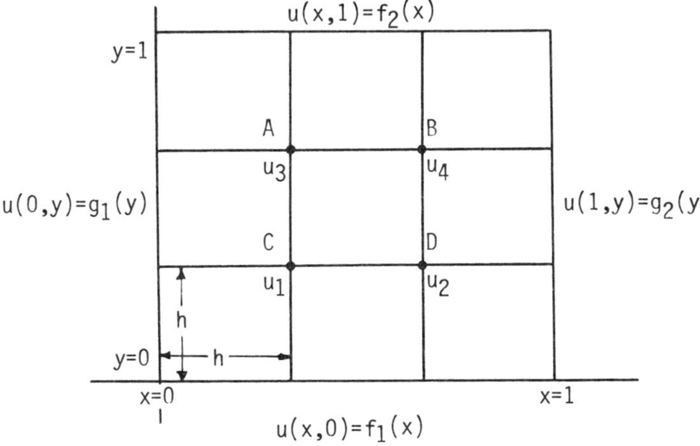

Figure 6.2.1. The region Γ and its covering mesh $\Gamma_{1/3}$.

At each point of Γ_h we replace the derivatives in equation (6.2.1) by central difference approximations with spacing h to give

$$\frac{[u(x+h,y)-2u(x,y)+u(x-h,y)]}{h^2} + \frac{[u(x,y+h)-2u(x,y)+u(x,y-h)]}{h^2} + E = 0, \qquad (6.2.2)$$

where $E=E(x,y,h)$ is the discretization error caused by such an approximation. Expanding each term in a Taylor series about (x,y) gives $E(x,y,h) \simeq h^2/12[\partial^4 u/\partial x^4 + \partial^4 u/\partial y^4]$ which tends to zero as $h \to 0$ provided u is sufficiently smooth. Therefore, for small values of h, we can approximate the differential equation at the point A by

$$u_4 - 2u_3 + g_1(2h) + f_2(h) - 2u_3 + u_1 = 0$$

or

$$u_1 \qquad - 4u_3 + \; u_4 = -g_1(2h) + f_2(h).$$

Similarly at B:

$$u_2 + \; u_3 - 4u_4 = -g_2(2h) - f_2(2h)$$

at C:

$$-4u_1 + u_2 + u_3 \qquad = -g_1(h) - f_1(h)$$

at D:

$$u_1 - 4u_2 \qquad + u_4 = -g_2(h) - f_1(2h)$$

which collectively may be written as

$$\begin{bmatrix} 1 & 0 & -4 & 1 \\ 0 & 1 & 1 & -4 \\ -4 & 1 & 1 & 0 \\ 1 & -4 & 0 & 1 \end{bmatrix} \begin{bmatrix} u_1 \\ u_2 \\ u_3 \\ u_4 \end{bmatrix} = \begin{bmatrix} -g_1(2h) - f_2(h) \\ -g_2(2h) - f_2(2h) \\ -g_1(h) - f_1(h) \\ -g_2(h) - f_1(2h) \end{bmatrix}.$$

These equations can then be solved to find values for u_1, u_2, u_3 and u_4. Notice that the matrix given above has no immediately apparent structure whereas if the equations are taken in the order C,D,B then A the corresponding equations become

$$\begin{bmatrix} -4 & 1 & 1 & 0 \\ 1 & -4 & 0 & 1 \\ 1 & 0 & -4 & 1 \\ 0 & 1 & 1 & -4 \end{bmatrix} \begin{bmatrix} u_1 \\ u_2 \\ u_3 \\ u_4 \end{bmatrix} = b, \qquad (6.2.3)$$

where the elements b are given by the right-hand sides of the previous equations taken in the order 3,4,1 and 2. This form is to be preferred because of the diagonal dominance exhibited by the matrix; it is then possible to determine the solution of these equations, without partial pivoting, as described in Chapter 1. Furthermore, the strict diagonal dominance of at least one row of the matrix guarantees that it is non-singular.

Example 6.2.1 Compute an approximate solution for Laplace's equation $u_{xx}+u_{yy}=0$ over the unit square subject to u vanishing at each point of the boundary except for the line segment $y=0$, $0\leqslant x\leqslant 1$ where $u(x,0)=\sin^2\pi x$.

Solution. For this problem $g_1=g_2=f_2=0$, $f_1=\sin^2\pi x$. Taking $h=1/3$ the right-hand side for equation (6.2.3) is $b=[-\sin^2\pi/3,-\sin^2 2\pi/3,0,0]$ which leads to the values $u_1=u_2=0.28125$, $u_3=u_4=0.09375$. The analytical solution of this problem is given in terms of an infinite series in Section 6.1. At the mesh points C and D this gives $u(h,h)=u(2h,h)=0.254526$, and at A and B $u(h,2h)=u(2h,2h)=0.0952523$. Considering the large value of h that has been used this gives reasonable agreement.

In general if the intervals for x and y are divided into N subintervals then there will be a total of $(N-1)^2$ internal points of the mesh. At each point the differential equation is approximated by a suitable difference approximation and collectively this gives a system of $(N-1)^2$ equations in $(N-1)^2$ unknown values. The form of these equations will depend upon the order in which the points of the mesh are numbered and the order in which the equations are taken and considerable research has been carried out to determine which is the optimal configuration. (See, for example, Varga (1962), Smith (1977).) Two particular orderings are of note. Firstly, the so called natural ordering in which the points are labelled 1 to N-1 along the first row of the mesh, N to 2N-2 along the second and so on. The second, called the red-black or chequerboard ordering, involves ordering the first and every alternate point along the first row, then the second and every alternate point of the second row, the first and every alternate point of the third row and so on, and then returning to label the remaining points. The natural ordering leads to a system of linear equations, the matrix of which has the form

$$\begin{bmatrix} A & I & 0 & 0 & . & . & 0 \\ I & A & I & 0 & . & . & 0 \\ 0 & I & A & I & . & . & 0 \\ . & . & . & . & . & . & . \\ . & . & . & 0 & I & A & I \\ 0 & . & . & . & 0 & I & A \end{bmatrix}, \qquad (6.2.4)$$

where each block is an $(N-1)\times(N-1)$ square matrix, the diagonal blocks being tridiagonal with -4 on the diagonal and 1 elsewhere, and I is the identity matrix of this order. (See equations (6.2.3) where each block is 2×2.) The chequerboard ordering leads to a matrix of the form

$$\left[\begin{array}{c|c} D_1 & B_1 \\ \hline B_2 & D_2 \end{array}\right], \qquad (6.2.5)$$

where D_1 and D_2 are diagonal matrices with diagonal elements -4 and B_1 and B_2 have unit entries on the main and 3 other diagonals.

Exercise 6.2.1 Determine the matrices associated with the difference approximation (6.2.2) with N=5 for the natural and chequerboard orderings.

The local error, $E(x,y,h)$, associated with equation (6.2.2) tends to zero as $h\to 0$ but this alone is insufficient to guarantee that at any point x,y of Γ the computed approximation converges to the solution at this point as h tends to zero; it is only sufficient to ensure that the difference between the approximation (6.2.2) and the differential equation tends to zero. In the next section it will be shown that as $h\to 0$ the computed approximation converges to the solution of the differential equation. Furthermore, if the solution is sufficiently smooth then the principal term in the error is $O(h^2)$ and so it is possible to extrapolate the computed solution.

Example 6.2.2 Using N=3 ($h={}^1/_3$) produces an approximation for the solution of Example 6.2.1 at $x=y={}^1/_3$ of 0.28125, and using N=6 ($h={}^1/_6$) gives 0.268090. Extrapolating these two results on the basis that the error is $O(h^2)$, see Section 1.5.3, gives an improved estimate of 0.254103 which compares well with the solution determined from the series given in Section 6.1 for which $u({}^1/_3,{}^1/_3)=0.254526$.

The previous example demonstrates that provided the solution is sufficiently smooth an accurate solution can be determined without resorting to higher order difference approximations which involve additional points of the mesh. (See, for example, Jain (1979)). In addition, the inclusion of additional nodes decreases the sparseness of the matrix and can lead to difficulties near the boundary of Γ. An alternative approach is to use the method of difference correction suggested by Fox. (See Section 3.4 for an example of the application of this approach as applied to a boundary value problem in ordinary differential equations.)

In general as the finite difference mesh is refined N grows larger and the number of equations increases rapidly. Although modern computers can cope with very large systems of equations there is a limit to the total space which is available. For example, if h=0.01 then N=100 and this gives a total of 99^2 unknown values in the mesh $\Gamma_{0.01}$. The number of elements in the matrix A is then 99^4 or 96,059,601, but an examination of the matrix shows that its elements are very sparsely distributed, for N=100 only about 0.05% of the elements are non-zero. However, the equations are of a banded form, for example they have bandwidth 2N-1 for the natural ordering. It is possible to solve these equations by storing only the non-zero diagonals using methods developed for just such equations; but this will still introduce elements on some of the diagonals which were originally empty. (See for example Meis and Marcowitz (1981)). This problem of "filling up" can be reduced to some extent by using orderings which attempt to minimise the number of non-zero entries in the final backward substitution associated with the Gaussian elimination (or LU decomposition). One such ordering due to George (1973) is based on successively dissecting the mesh into smaller regions, for example a mesh of dimension 5×5 might be ordered according to Figure 6.2.2.

1	5	17	7	2
6	13	21	14	8
18	22	25	23	19
9	15	24	16	12
3	10	20	11	4

Figure 6.2.2. George's Successive Dissection Ordering.

A summary of other possible orderings and their respective advantages is given by Fox (1977). It is also possible to take advantage of the sparseness by storing only the non-zero entries, and their positions, and by doing so reduce the overall storage requirements. However, careful track must be kept not only of the positions of the original non-zero entries but also any additional entries which are created by the Gaussian elimination or LU decomposition which is used to solve the equations. (See Ortega and Poole (1981), Erisman and Reid (1981) and Reid (1977).). Another approach is to attempt to solve the equations by an iterative process. The solution of a full system of M linear equations by the direct elimination process involves approximately $O(M^3)$ multiplications whilst most iterative methods involve $O(M^2)$ per iteration. Therefore, if a satisfactory solution can be found in less than O(M) steps iterative methods can be very useful. Against this must be weighed the practical problems associated with iterative methods, for example, the difficulty of specifying a suitable termination criterion, the growth of rounding errors involved with any repetitive process and the determination of a suitable initial estimate for the solution. However, provided the matrix is sparse the advantage to be gained for large systems by using an iterative method can be worthwhile. The application of either Jacobi's method or the Gauss-Seidel iteration is straightforward though the SOR method is to be preferred if it is possible to obtain a good estimate for the acceleration parameter ω. (See Section 1.5). Unfortunately, there is no known method for determining such a value for a given arbitrary matrix. However, if the matrix has certain properties then it is possible to find the optimum acceleration parameter. For example, if a matrix is block tridiagonal, i.e. it is composed of 3 diagonals of blocks where the blocks lying on the main diagonal are themselves diagonal matrices, and the Jacobi method converges then the optimum value for ω is given by

$$\omega_{opt} = 2/(1 + \sqrt{1-\rho(G_J)^2}), \qquad (6.2.6)$$

where $\rho(G_J)$ is the spectral radius of the associated Jacobi iterative process. Even then the application of this result is limited by being able to determine $\rho(G_J)$. The matrix (6.2.5) is block tridiagonal but (6.2.4) is not; but it can be transformed to (6.2.5) by carrying out a finite sequence of row and column permutations in which case it is said to possess property A. Further details concerning optimal SOR parameters are given by Varga (1962) and Young (1971).

To illustrate how iterative methods may be used to solve equations of the form (6.2.1) we consider the problem given in example 6.2.1 for which the unit square which has be covered by a mesh with spacing $h={}^1/_N$. Writing out the difference equations corresponding to equation (6.2.2) gives a matrix system of equations which is of dimension $(N-1)^2$. However, it is possible to avoid the explicit evaluation of this large matrix by the following procedure. If U_{mn} is the computed approximation for the solution $u(mh,nh)$ then it may be determined by applying the following simple algorithm:-

(1) for m = 1 to N-1

(2) for n = 1 to N-1

(3) replace U_{mn} by $(U_{m+1,n}+U_{m-1,n}+U_{m,n+1}+U_{m,n-1})/4$

(4) repeat

It is of course possible to replace the natural ordering used in this algorithm by any other suitable arrangement. This is equivalent to applying the Gauss-Seidel method to the system of equations generated by using (6.2.2) at each point of the mesh. This algorithm can be adapted to produce a simple SOR iteration by replacing step (3) by

(3)i set $s = (U_{m+1,n}+U_{m-1,n}+U_{m,n+1}+U_{m,n-1})/4$
(3)ii replace U_{mn} by $U_{mn} + \omega[s - U_{mn}]$.

The determination of the optimal SOR acceleration parameter for this problem involves the calculation of the spectral radius of the associated Gauss-Seidel method for it is possible to show that this is equal to $\sqrt{\rho(G_J)}$ from which ω_{opt} can be determined by equation (6.2.6). In practice this is not always the case and it may be necessary to resort to more empirical methods. For example, by using a variety of values for ω and a crude termination criterion it is possible to estimate the optimum value for ω, this value is then used in an SOR iteration with a finer termination criterion.

Example 6.2.3 Consider the solution of $u_{xx}+u_{yy}=0$, (x,y) in $\Gamma=\{0<x,y<1\}$, subject to $u=0$ on the boundary except along the line $y=0$, $0\leqslant x\leqslant 1$, where $u(x,y)=\sin^2\pi x$. Taking $N=10$, gives $h=0.1$ and a mesh of 81 points for which the Gauss-Seidel method, iterating until the maximum difference between successive elements of the solution differs by less than 10^{-7}, converges in 101 iterations. Using the following values for the SOR parameter ω and a termination criterion of 10^{-3} produces convergence in the following number of iterations.

ω	1	1.2	1.3	1.4	1.5	1.6	1.7
Number of Iterations	32	24	20	16	13	13	21

It would appear that the optimal value for ω is between 1.5 and 1.6. Taking $\omega=1.55$ the SOR process converges, such that successive approximate values differ by less than 10^{-7}, in 24 iterations. As this set of equations can be transformed to a block tridiagonal matrix by permuting rows and corresponding columns the optimal SOR parameter is given by (6.2.6). Furthermore, the spectral radius of the Gauss-Seidel procedure can be estimated from the Lyusternik acceleration process, which gives $\rho(G_{GS}) \simeq [\rho(G_J)]^2 \simeq 0.90488$ and so $\omega_{opt} \simeq 2/(1+\sqrt{1-0.90488}) \simeq 1.53$. The amount of work involved to solve this problem could be reduced by noting that the solution must be symmetric about the line x=0.5, therefore, u(0.5-x,y)=u(0.5+x,y) which is equivalent to setting $U_{5-m,n}=U_{5+m,n}$, for m=1,2,3,4 which reduces the problem to a system of 45 equations.

A simple alternative between the direct and iterative methods is given by block iterative procedures. As an example we first write out the equations corresponding to applying the difference approximation (6.2.3) at all the points along the first line of the mesh, keeping all other values fixed. These equations are tridiagonal and are easily solved by a direct method. This process is then carried out for the second line of the mesh and so on until all points of the mesh have been updated. We then return to the first line and repeat the process. This procedure can be improved by alternately solving parallel to the x axis on one sweep and then parallel to the y axis on the next. This gives the so called Alternating Direction Implicit (ADI) methods which are equivalent to those described in Section 4.6 where a solution to the equation $u_t=\sigma(u_{xx}+u_{yy})$ was sought, here we are looking for the steady state solution of the same problem. A summary of some of the most useful ADI results is given by Ames (1977).

Exercise 6.2.2 Determine a solution for Laplace's equation on the unit square subject to the solution vanishing at all points of the boundary except along the line y=0, $0 \leq x \leq 1$ where the solution takes the values $\sin^2 \pi x$. Use mesh sizes h=0.1 and h=0.05 and extrapolate the computed approximations to produce a more accurate solution. Estimate the discretization error at each point of the mesh and then compare the extrapolated solution with the analytical solution given in Section 6.1.

6.3 Generalisations

The inhomogeneous form of Laplace's equation,

$$u_{xx} + u_{yy} = f, \qquad (x,y) \text{ in } \Gamma \subset R^2, \tag{6.3.1}$$

is called Poisson's equation. We now consider the solution of this equation on a simply connected region Γ for which u is specified on the boundary $\partial\Gamma$. (A simply connected region is such that any curve lying within the region can be shrunk to a point in such a way that it always lies entirely within the region.). We begin by assuming that the shape of Γ is such that the mesh lines intersect $\partial\Gamma$ only at points of the mesh.

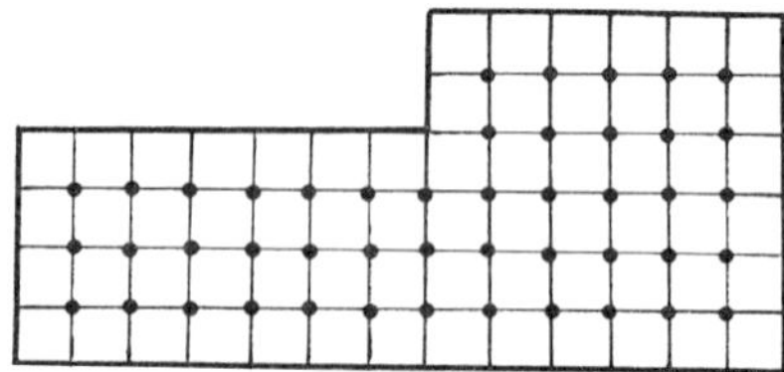

Figure 6.3.1. A **region** Γ **and** a **covering mesh** Γ_h.

At each internal mesh point the differential equation (6.3.1) is replaced by the difference approximation

$$L_h U_{mn} = [U_{m+1,n} + U_{m-1,n} + U_{m,n-1} + U_{m,n-1} - 4U_{mn}]/h^2 = f_{mn}, \tag{6.3.2}$$

then

$$L_h u(mh,nh) = f_{mn} - E_{mn}, \tag{6.3.3}$$

where E_{mn} is the local discretization error at the point x=mh, y=nh. Subtracting (6.3.2) from (6.3.3) the error $e_{mn} = u(mh,nh) - U_{mn}$ satisfies

$$L_h e_{mn} = L_h\{u(mh,nh) - U_{mn}\} = -E_{mn}.$$

We now show that $\max|e_{mn}| \leq {}^{h^2}/_{24}\, M_4$, where M_4 is an upper bound for $|{}^{\partial^4 u}/_{\partial x^4}|$ and $|{}^{\partial^4 u}/_{\partial y^4}|$. Define $w^{\pm}_{mn}=\pm e_{mn}+R\phi_{mn}$ where $\phi_{mn}=(m^2+n^2)h^2/4$ then $L_h w^{\pm}_{mn}=\pm L_h e_{mn}+RL\phi_{mn}=\pm L_h e_{mn}+R$ since $L_h\phi_{mn}=1$. Now select $R=\max|L_h e_{mn}|\leq h^2M_4/_6$ and then $L_h w^{\pm}_{mn}\geq 0$ for all points in Γ. From this we show that

$$\max_{\Gamma} w^{\pm}_{mn} \leq \max_{\partial\Gamma} w^{\pm}_{mn}$$

which is called a maximum principle. If $w^{\pm}_{mn}$ has a maximum M within Γ at some point P for which m=m' and n=n' then $M=w^{\pm}_{m'n'}\geq w^{\pm}_{m'\pm 1,n'\pm 1}$ for all points in the interior of Γ and $M>w^{\pm}_{mn}$ for (mh,nh) on the boundary $\partial\Gamma$. However, $L_h w^{\pm}_{m'n'}\geq 0$, therefore, each of $w^{\pm}_{m'\pm 1,n'\pm 1}$ is also equal to M. Repeating this argument for any of the points $m'\pm 1,n'\pm 1$ it follows that $w^{\pm}_{mn}=M$ for all points in Γ_h and $\partial\Gamma_h$ which contradicts the assumption that $w^{\pm}_{mn}<M$ for points on the boundary, therefore, we have the required result. Since $w^{\pm}_{mn}=\pm e_{mn}+R\phi_{mn}$ and $R\phi_{mn}\geq 0$ then $\pm e_{mn}\leq w^{\pm}_{mn}$ from which we conclude that

$$\max_{\Gamma}|e_{mn}| \leq \max_{\partial\Gamma}|e_{mn}| + {}^1/_2\max_{\partial\Gamma}|L_h e_{mn}| \leq {}^{h^2}/_{24}\, M_4$$

as required. Therefore, as $h\to 0$ the difference solution converges to the the solution of the differential equation.

The extension to the variable coefficient elliptic partial differential equation

$$a(x,y)u_{xx} + c(x,y)u_{yy} = F(x,y), \quad (x,y) \text{ in } \Gamma,$$

is straightforward. At each point of a suitable mesh Γ_h the differential equation is approximated by

$$a(mh,nh)\frac{\delta_x^2 U_{mn}}{h^2} + c(mh,nh)\frac{\delta_y^2 U_{mn}}{h^2} = F(mh,nh).$$

As before this produces a system of linear equations which has the same dimension as the number of points in the mesh Γ_h and which may then be solved to find values for U_{mn}.

The restriction that the mesh intersects the boundary only at mesh points may be a severe limitation in an irregular region where it may not be possible to arrange a uniform mesh and yet satisfy this constraint. Let us assume that near a boundary the mesh points are distributed as shown on Figure 6.3.2.

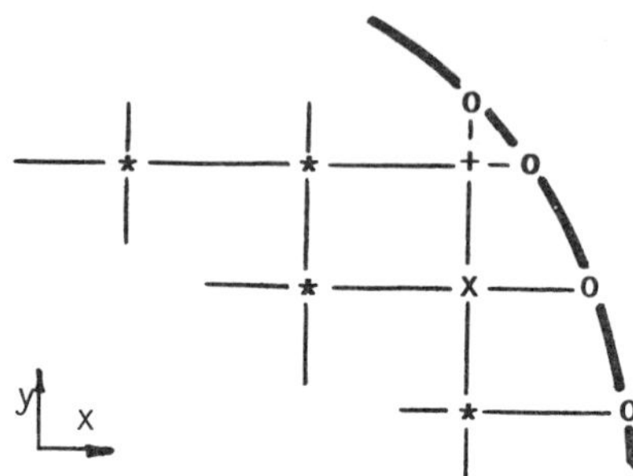

Figure 6.3.2. Mesh points near an irregular boundary.

* - internal nodes. o - boundary points.

+ and x - boundary nodes.

The points denoted * are called internal nodes or pivots if they are surrounded by mesh points within Γ and at each point of this type the difference equation corresponding to (6.3.2) can be applied. The remaining points within Γ are called boundary nodes or pivots. At a boundary node it is necessary to replace one or both of u_{xx} and u_{yy} by uncentred differences. For example, at the point x u_{yy} can be replaced by a central difference but not u_{xx}. At the point + both derivatives need to be replaced by uncentred differences. Consider the situation given in Figure 6.3.3 for which

$$u_{xx} = \frac{2}{h^2}\left[\frac{1}{(\alpha+\beta)} \left(\frac{U_1}{\beta}+\frac{U_{-1}}{\alpha}\right) - \frac{U_0}{\alpha\beta}\right] + O(h). \qquad (6.3.4)$$

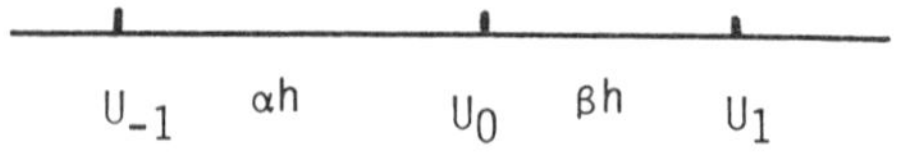

U_{-1} αh U_0 βh U_1

Figure 6.3.3 An uncentred difference approximation for u_{xx}.

The uncentred difference approximation given by equation (6.3.4) reduces to the normal centred approximation when $\alpha=\beta$. The derivative

u_{yy} can be replaced in a similar way. Notice that this process is $O(h)$ when $\alpha \neq \beta$ which will detract from the overall $O(h^2)$ of the difference approximation (6.2.2).

An alternative to using uncentred differences is to make a change of variable such that the resulting problem is defined on a region which can be covered by a mesh which intersects the boundary only at mesh points.

Example 6.3.1 Consider the solution of Laplace's equation $u_{xx}+u_{yy}=0$ on the semi-circular region $\Gamma=\{ x^2+y^2 \leqslant 1, y \geqslant 0 \}$ subject to the boundary conditions $u=x^2+y^2$ for (x,y) in $\partial\Gamma$. The change of variable $x=r\cos\theta$, $y=r\sin\theta$ produces Laplace's equation in polar co-ordinates, i.e.

$$\partial^2 u/\partial r^2 + 1/r \; \partial u/\partial r + 1/r^2 \; \partial^2 u/\partial \theta^2 = 0.$$

The region $0 \leqslant r \leqslant 1$, $0 \leqslant \theta \leqslant \pi$, is now covered by a mesh made up of the semi-circles $r=m\delta r$, $(m=1,2,..,M)$ and straight lines $\theta=n\delta\theta$, $n=0,1,..,N$, where $M\delta r=1$ and $N\delta\theta=\pi$, as shown in Figure 6.3.4.

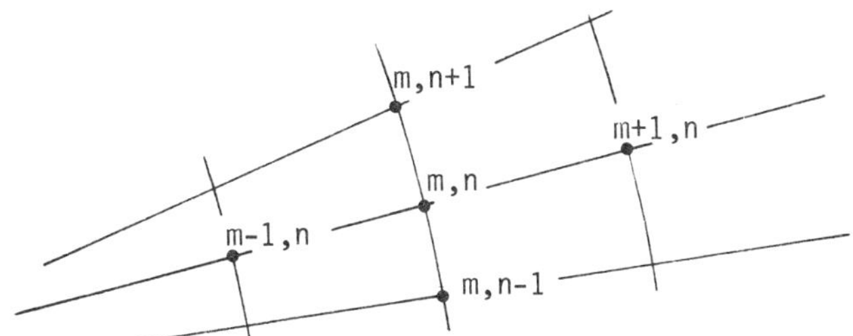

Figure 6.3.4 The covering mesh in polar co-ordinates.

The differential equation is then approximated by

$$\frac{(U_{m+1,n}-2U_{mn}+U_{m-1,n})}{\delta r^2} + \frac{1}{m\delta r}\frac{[U_{m+1,n}-U_{m-1,n}]}{2\delta r} + \frac{1}{(m\delta r)^2}\frac{(U_{m,n+1}-2U_{mn}+U_{m,n-1})}{\delta\theta^2}=0,$$

where $U_{mn} \simeq u(m\delta r, n\delta\theta)$, which has error $O(\delta r^2, \delta\theta^2)$. This gives a system of linear equations which may be solved for the unknown mesh values. This problem can be simplified by using the symmetry $x \to -x$ and so it is only necessary to solve this problem for $0 \leq r \leq 1$, $0 \leq \theta \leq \pi/2$.

The use of non-square meshes can be further extended by using other meshes which are based on tessellations other than squares or rectangles. To illustrate this we consider the following example.

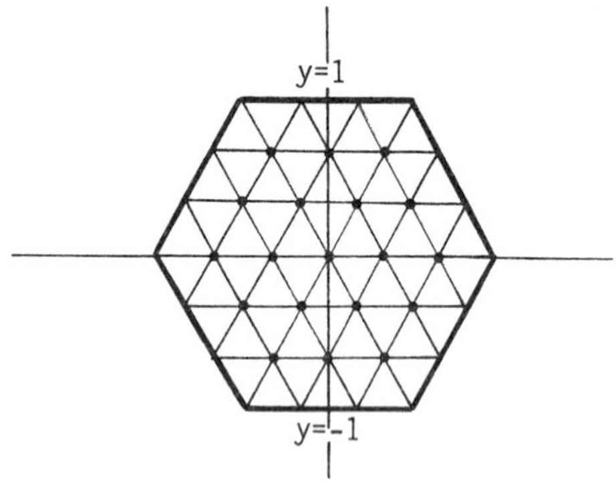

Example 6.3.2 Consider the solution of Laplace's equation in the hexagonal region shown above and subject to u(x,y) being given on the boundary. This region may be covered by a mesh made up by the points of intersection of three families of curves which are drawn parallel to the sides of the hexagon. A typical mesh point is now surrounded by 6 neighbouring points.

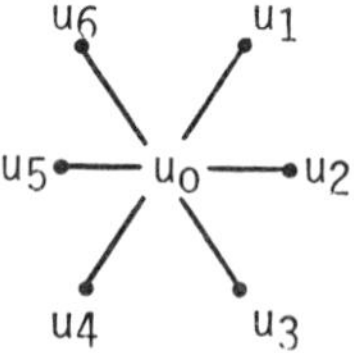

By expanding each term in a Taylor series about the centre it is straightforward to show that

$$u_{xx} + u_{yy} = \{ \sum_{i=1..6} U_i - 6U_0 \}/h^2 + O(h^2).$$

Exercise 6.3.1 Verify the above expression and show that it leads to a diagonally dominant system of equations.

Exercise 6.3.2 Use the three-fold symmetry of the above example to reduce the number of equations which need to be solved in order to find a solution for Example 6.3.2.

6.4 Neumann Problems

The solution of Poisson's equation $u_{xx}+u_{yy}=f$, for (x,y) in $\Gamma=\{0\leqslant x,y\leqslant 1\}$, subject to the boundary conditions

$$\partial u/\partial n = \begin{cases} f_1(x), & y=0,\ 0\leqslant x\leqslant 1, \\ f_2(x), & y=1,\ 0\leqslant x\leqslant 1, \\ g_1(y), & x=0,\ 0\leqslant y\leqslant 1, \\ g_2(y), & x=1,\ 0\leqslant y\leqslant 1. \end{cases} \qquad (6.4.1)$$

where $\partial u/\partial n$ is the derivative of u along the outward normal direction to the boundary $\partial\Gamma$ is called a **Neumann problem.** The solution of this problem is not unique, for if u is a solution then so is u+c, where c is any arbitrary constant. In order to obtain a unique solution it is necessary to specify an additional condition. This may take the form of specifying the solution at just one point or perhaps involve an integral of u or $\partial u/\partial n$ around the boundary.

The region Γ is covered with a mesh Γ_h where Nh=1 and then the difference approximation corresponding to equation (6.3.2) is applied at each internal node. This gives a total of $(N-1)^2$ equations in all. At each point of the boundary it is possible to approximate $\partial u/\partial n$ involving only one point on the boundary and its closest internal node, denoted by "*" and "o" respectively in Figure 6.4.1, but this gives an O(h) approximation. Alternatively, spurious points are introduced along the entire length of the boundary as shown in Figure 6.4.1.

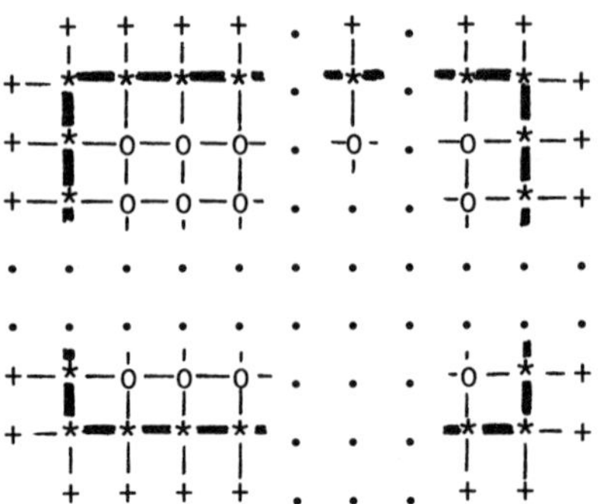

Figure 6.4.1. Spurious points (+) introduced for Neumann problems.
+ - spurious points, * - boundary points, o - internal nodes.

The region upon which a solution is required is bounded by the points marked "*". At each of the points denoted by "o" the difference approximation (6.3.2) can be applied. The boundary conditions (6.4.1) are then replaced by

$$\begin{aligned} U_{-1,n} - U_{1,n} &= 2hg_1(nh), & 0 \leq n \leq N, \\ U_{N+1,n} - U_{N-1,n} &= 2hg_2(nh), & \\ U_{m,-1} - U_{m,1} &= 2hf_1(mh), & 0 \leq m \leq N, \\ U_{m,N+1} - U_{m,N-1} &= 2hf_2(mh) & \end{aligned}$$

and then the differential equation is applied at the points *. This gives a total of $(N+3)^2-4$ equations in $(N+3)^2-4$ unknowns. However, it is possible to eliminate the spurious points beforehand to reduce the problem to one in $(N+1)^2$ unknowns which must also satisfy an additional condition to ensure a unique solution. This process is analogous to the solution of an ordinary differential equation subject to derivative boundary conditions as described in Example 3.4.1.

If the boundary of Γ, $\partial\Gamma$, is curved then it is necessary to extend the process described previously. Let us assume that $\partial u/\partial n = f(x,y)$, then we may then approximate the normal derivative at Q by

$$\partial u/\partial n|_Q \simeq [U_Q - U_S]/\beta h \simeq [U_Q - (1-\alpha)U_T + \alpha U_P]/\beta h. \qquad (6.4.2)$$

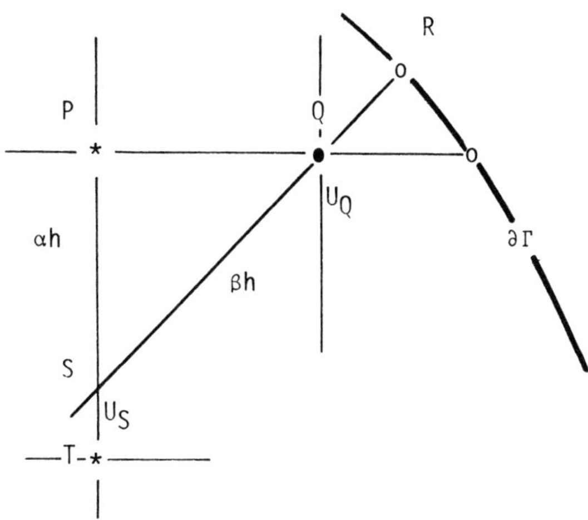

Exercise 6.4.1 Consider the solution of Laplace's equation subject to the boundary conditions that $^{\partial u}/_{\partial n}=0$ on the curved boundary $x^2+y^2=1$ and u=0 for y=0. Determine an approximate solution using a square cartesian mesh with spacing h=0.25. Consider the same problem after a change to polar co-ordinates taking $\delta r=0.25$, $\delta\theta={}^{\pi}/_{6}$. Use the symmetry of the problem about the line x=0, $(\theta={}^{\pi}/_{2})$, to reduce the overall amount of computation involved.

6.5 Singularities

All difference approximations have been derived on the assumption that the solution is sufficiently smooth to allow us to expand it in a Taylor series about a specified point and then examine the local discretization error; however, this is not always the case. There are two possible causes of such singularities, the first involves one or more terms of the differential equation becoming undefined, the second is a little less obvious and involves difficulties at or near the boundaries.

Example 6.5.1 Laplace's equation in polar co-ordinates is given by

$$\partial^2u/\partial r^2 + 1/r\ \partial u/\partial r + 1/r^2\ \partial^2u/\partial\theta^2 = 0.$$

As $r\to 0$ there appears to be a singularity in the coefficients but this may be overcome as follows. At $r=0$ we revert to approximating Laplace's equation in cartesian co-ordinates and construct a circle of radius δr about the origin which cuts the x and y axes at the points $(\pm\delta r,\pm\delta r)$ where the approximate solution takes the values $U_{\pm 1,\pm 1}$. At $x=y=0$ we may approximate Laplace's equation by

$$[U_{1,1}+U_{1,-1}+U_{-1,1}+U_{-1,-1}-4U_{0,0}]/\delta r^2 = u_{xx}+u_{yy} + O(\delta r^2),$$

where $U_{0,0}$ is a computed approximation for the solution at the origin. Therefore, $U_{0,0}=1/4\ [U_{1,1}+U_{1,-1}+U_{-1,1}+U_{-1,-1}]$. However, rotating the axes by a small amount a similar result will hold, therefore, we can approximate $U_{0,0}$ by averaging the computed solution at all the points of the mesh which intersect a circle of radius δr.

The technique given in Example 6.5.1 extends to Poisson's equation but the situation is further complicated by possible singularities in the inhomogeneous term, $f(x,y)$, in equation (6.3.1). Hubbard (1966) has shown that if the solution $u(x,y)$ is finite and that its first j derivatives, $j=1,2,3,4$, are bounded by a term $O(\varepsilon^{\alpha-j+1})$, where ε is the distance from the point (x,y) to the singularity, then the error $e_{mn}=u(mh,nh)-U_{mn}$ is $O(h^\alpha+h^2)$ provided that the solution at the boundary is determined according to equation (6.4.2).

We now consider a second possible source of singularities in the solution of a given differential equation. It has been assumed that any boundary conditions which the differential equation satisfies change smoothly, if not then singularities may occur in the solution. For example, with equation (6.2.1) the boundary conditions at the corners of the unit square should change smoothly, i.e. at x=y=0

$$\lim_{x\to 0} f_1(x) = \lim_{y\to 0} g_1(y).$$

If this is not the case then it is possible to replace the boundary condition at x=0, y=0 by $[f_1(0)+g_1(0)]/2$. Unlike hyperbolic equations such a discontinuity cannot propagate into the region where the solution is required as there are no real characteristic directions along which it can travel and so the discontinuity will only contaminate a very small region about the point (0,0). A similar result holds at each of the other 3 corners, however, if the boundary is made up of a number of piecewise smooth curves any two of which meet with an interior angle π/α, where $1/2 \leqslant \alpha < \infty$, then it is possible to show that the error between the computed solution and its correct counterpart is proportional to a term $O(h^{\alpha}+h^2)$ provided the jth derivative of the solution is bounded by terms of $O(\varepsilon^{\alpha-j})$, $j=0,1,2,\ldots$. (See Hubbard (1966)). The worst possible outcome is that the error is $O(\sqrt{h})$ when the segments of the boundary meet at a very acute exterior angle. This difficulty, usually referred to as the problem of a re-entrant corner, can be overcome by selecting a finer mesh near such points or by attempting to deal with the singularity analytically. There are two main approaches to doing this. The first involves attempting to find an analytic solution valid near the corner. The second involves a mapping of the region in which the solution is required onto one without re-entrant corners, i.e. where a sudden change in the direction of the boundary at a point becomes a smooth change along a curve in the region bounded by the new independent variables.

Example 6.5.2 Consider the solution of Laplace's equation in the region ABCD, subject to the given boundary conditions and $u^+_y=u^-_y=0$ on FE. (The ± denote one sided derivatives in the respective directions).

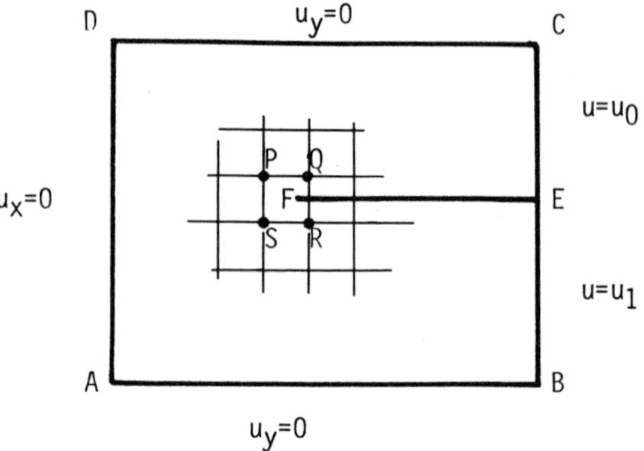

Near the point F the solution may be singular and so we attempt to construct an analytic approximation valid near this point. The functions $r^k\cos k\theta$ and $r^k\sin k\theta$ satisfy Laplace's equation for all values of k, therefore, as a possible solution near the point F, where the boundary has an internal angle α, we consider a linear combination of such terms, i.e.

$$u(r,\theta)= \sum_{n=1..\infty} r^{n\pi/\alpha}\,[a_n\sin{}^{n\pi\theta}/_{\alpha} + b_n\cos{}^{n\pi\theta}/_{\alpha}]. \qquad (6.5.1)$$

For this example $a_n=0$, for all n, since $u_y=0$ on FE. Therefore, using the first 4 terms gives an analytic approximation for the solution near the point F

$$u_4(r,\theta) = b_0 + b_1r^{1/2}\cos{}^1/_2\theta + b_2r\cos\theta + b_3r^{3/2}\cos{}^3/_2\theta,$$

since $\alpha=2\pi$. To update an approximate solution we apply the normal five point difference scheme at each point of the mesh, including P,Q,R and S, but not at F. Finally, to update the solution at F we fit the above function at the points P,Q,R and S and use this as a local solution near F. This process is then iterated to convergence.

Example 6.5.3 The change of variables $\xi=x^ay^b$, $\eta=y^c$ can often be beneficial in stretching a singular point to give a line along which there is a smooth change in any boundary condition.

Exercise 6.5.1 Show that the change of variables $\xi = x/\sqrt{y}$, $\eta = 1/\sqrt{y}$ transforms the positive x and y axes to give (i) the line $\xi=0$ corresponding to $x=0$, $y>0$, (ii) the line $\eta=0$ corresponding the the origin $x=y=0$ and (iii) the point at infinity for the line $y=0$, $x>0$. Under this transformation the equation $u_y=u_{xx}$ becomes $u_{\xi\xi}+{}^1/_2\xi u_\xi={}^1/_2\eta u_\eta$. Find the equation corresponding to Laplace's equation in the ξ-η plane.

6.6 Finite Element Methods

The solution of the elliptic partial differential equation

$$\partial/\partial x(p\partial u/\partial x) + \partial/\partial y(q\partial u/\partial y) + ru = f(x,y), \quad (x,y) \text{ in } \Gamma \qquad (6.6.1)$$

subject to u being specified on a part, $\partial\Gamma_1$, of the boundary of Γ, and $(pu_x,qu_y)\cdot n+g_1(x,y)u=g_2(x,y)$ on the rest, $\partial\Gamma_2$, where $p(x,y)$, $q(x,y)$ and $g_1(x,y) > 0$ and $r(x,y)\leq 0$, is equivalent to finding a stationary point for the functional

$$I(u) = \int_\Gamma {}^1/_2[p(u_x)^2 + q(u_y)^2 - ru^2]\, d\Gamma + \int_{\partial\Gamma_2} [-g_2u + {}^1/_2g_1u^2]\, dS. \qquad (6.6.2).$$

(A simple derivation of this result for Laplace's equation subject to Dirichlet boundary conditions is given by Elsgolc (1963), a complete derivation is given by Strang and Fix (1973)). For example, if $u(\mathbf{c})=u(c_1,c_2,\ldots,c_N)$ is a trial solution for either equation (6.6.1) or (6.6.2) then by selecting the parameters $\{c_i\}$, $i=1,..,N$, so that $I=I(c_1,c_2,\ldots,c_N)$ has a stationary value i.e. $\partial I/\partial c_i=0$, $i=1,\ldots,N$, gives a system of N equations in the unknown parameters. The solution of these equations will give the best approximation amongst the set of functions generated by taking all possible values of the parameters. This was the original basis of the Finite Element Method but it also possible to consider differential equations for which there is no directly related variational principle.

In order to consider the finite element method in a more general setting we introduce the following notation. Let Ω be the set of functions which map $R^2 \to R$ and L and B be differential operators which map $\Omega \to R^2$ then equation (6.6.1) may be written as Lu=f, for values of $\mathbf{x}=(x,y)$ in Γ, where $L={}^{\partial}/_{\partial x} p {}^{\partial}/_{\partial x} + {}^{\partial}/_{\partial y} q {}^{\partial}/_{\partial y} + r$, subject to Bu=g for x in $\partial\Gamma$. As an approximate solution of this problem consider the trial function

$$w_N(\mathbf{x},\mathbf{c}) = \phi_0(\mathbf{x}) + \sum_{j=1..N} c_j \, \phi_j(\mathbf{x}), \qquad (6.6.3)$$

where the basis function ϕ_j are selected such that $B\phi_0=g$ and $B\phi_j=0$, $j>1$, then w_N satisfies the boundary conditions. The error

$$E(\mathbf{x},\mathbf{c}) = Lw_N - f \qquad (6.6.4)$$

is a measure of how close the trial solution (6.6.3) is to the solution of the differential equation Lu=f and should w_N tend to the solution u as $N\to\infty$. For example, in Section 6.1 the trial functions could have been selected as $\phi_j(x,y)=\sinh j\pi(1-y).\sin j\pi x$. The way in which the coefficients $\{c_j\}$, j=1,..,N, are selected gives the various finite element techniques. For example, for a particular choice of basis functions the coefficients $\{c_j\}$, j=1,..,N, could be chosen such that

$$\int_\Gamma f_i E \, d\Gamma = 0, \quad i=1,..,N, \qquad (6.6.5)$$

where the functions $f_i(\mathbf{x})$ are suitable weight functions. Once the functions f_i are specified equation (6.6.5) gives a system of N equations for determining the coefficients $\{c_j\}$, j=1,..,N. This is called the **Weighted Residual Method** (WRM). One possible choice of weight functions is to take $f_j=\phi_j$, j=1,..,N, which gives the **Bubnov-Galerkin** method. An alternative choice is to set $f_j={}^{\partial E}/_{\partial c_j}$, j=1,..,N, then the functional $J(u)={}^1/_2\int_\Gamma E^2 d\Gamma$ is minimized when ${}^{\partial J}/_{\partial c_i}=\int_\Gamma E \, {}^{\partial E}/_{\partial c_i} d\Gamma=0$; this gives the **Least Squares Method.**

The finite element method in the case of partial differential equations is very similar to the technique used for ordinary differential equations. The result of applying this method to a given differential equation is to produce a system of discrete equations which, when solved, produce an approximate solution. An example of its application to an ordinary differential equation is given in Section 3.5.

Example 6.6.1 Consider the solution of Poisson's equation $u_{xx}+u_{yy}=f$, for (x,y) in a region Γ, subject to $u=g$, for points on the boundary $\partial\Gamma$, where Γ is given in Figure 6.6.1. The region Γ is covered by triangular elements and then an approximate value for the solution at each vertex is determined.

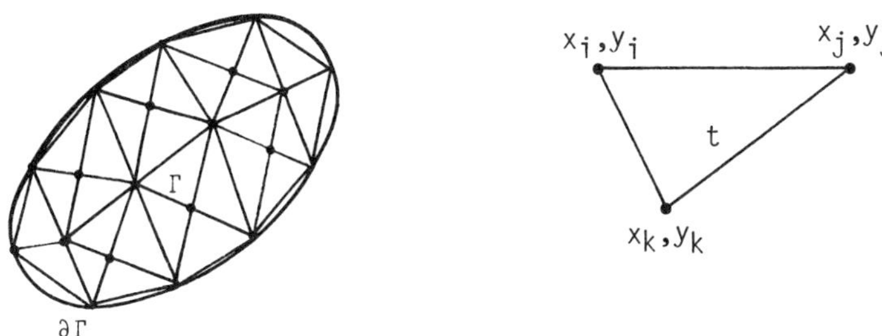

Figure 6.6.1 A finite element division of Γ for Example 6.6.1.

Notice that the triangles can be selected so that any point of intersection with the boundary is an apex of at least 1 triangle and so the shape of the region is no longer the problem it was with finite difference methods in Section 6.4. Let us assume that in each triangular region it is possible to approximate the solution by

$$u(x,y) = t_1 + t_2x + t_3y, \tag{6.6.6}$$

then for a particular triangle, t, with vertices (x_p,y_p) at which the solution takes values U_p, $p=i,j,k$, therefore,

$$U_p = t_1 + t_2x_p + t_3y_p, \qquad p=i,j,k. \tag{6.6.7}$$

Equations (6.6.7) may be solved for t_1, t_2 and t_3. Alternatively, define

$$t_i(x,y)=[(x_jy_k-x_ky_j)+(y_j-y_k)x+(x_j-x_k)y]/_\Delta t,$$

where Δ^t is the area of the triangular element t, and similarly $t_j(x,y)$ and $t_k(x,y)$ by permutation of i, j and k. The functions $t_p(x,y)$, p=i,j,k, are called shape functions for this element; t_p is 1 at (x_p,y_p) and vanishes at the other vertices therefore, in triangle t

$$u(x,y) = t_i(x,y)U_i + t_j(x,y)U_j + t_k(x,y)U_k. \qquad (6.6.8)$$

Each node is involved not only in the triangle t but also any other element with a vertex at this point, therefore, we define the basis functions

$$\phi p(x,y) = \begin{cases} t_p(x,y) \text{ if } (x,y) \text{ is in triangle t which has p a vertex,} \\ 0, \qquad \text{otherwise,} \end{cases} \qquad (6.6.9)$$

and then consider the trial solution

$$u_N(x,y) = \sum_{i=1..N} U_i\phi_i. \qquad (6.6.10)$$

The accuracy of the method can be improved by using a higher order approximation in each element, for example, (6.6.6) could be replaced by

$$u(x,y) = t_1 + t_2x + t_3y + t_4xy + t_5x^2 + t_6y^2,$$

but to find the shape functions for this expression would require knowledge of the solution at points other than the vertices of the triangle. For example, in addition to the vertices of the element the solution could be matched at the mid-points of its sides. An alternative way of producing a more accurate solution is to use different shaped elements.

Exercise 6.6.1 Show that the coefficients t_p, p=i,j,k, in equation (6.6.7) are given by

$$t_1=[(x_jy_k-x_ky_j)U_i + (x_ky_i-x_iy_k)U_j + (x_iy_j-x_jy_i)U_k]/\Delta,$$
$$t_2=[(y_j-y_k)U_i + (y_k-y_i)U_j + (y_i-y_j)U_k]/\Delta,$$
$$t_3=[(x_j-x_k)U_i + (x_k-x_i)U_j + (x_i-x_j)U_k]/\Delta,$$

where Δ is the area of the triangular element bounded by (x_p,y_p), p=i,j,k. Hence, show that (6.6.6) and (6.6.8) are equivalent.

Exercise 6.6.2 If rectangular elements, r, are used then each element will have 4 vertices and the natural choice of trial function is

$$u(x,y)=r_1 + r_2x + r_3y + r_4xy.$$

If the vertices are (x_p,y_p), p=i,j,k,l, labelled anti-clockwise around the the rectangle, show that the shape functions are given by

$$r_i=(x-x_j)(y-y_l)/[(x_i-x_j)(y_i-y_l)],$$

and similarly for the vertices j,k and l.

Once the form of the approximate trial solution is determined it is substituted into a suitable functional and a stationary value determined.

Example 6.6.2 The solution of Poisson's equation in a region Γ, subject to the Dirichlet boundary condition that u is specified on the boundary, is equivalent to finding a stationary value of the functional

$$I(u) = \iint_\Gamma [{}^1/_2(u_x{}^2 + u_y{}^2) + fu)dxdy.$$

For $u_N(x,y)=\sum U_i\phi_i(x,y)$, a stationary value is given by $\partial I/\partial U_i=0$, i=1,....,N. The solution of this system of N equations determines the coefficients U_i, and hence the solution (6.6.10). Because of the way in which the basis functions have been defined not every element will make a contribution to each of these equations; contributions come only from those elements which have a vertex at (x_i,y_i). Therefore, we write

$$\partial I/\partial U_i = \sum \partial I_t/\partial U_i,$$

where the summation is over all the elements which have a vertex at (x_i,y_i).

Consider a triangular region which has vertices (x_p,y_p), $p=i,j,k$, then the contribution from this element is

$$I_t(U_N)=\iint_t \{ 1/2[\partial/\partial x(t_iU_i+t_jU_j+t_kU_k)^2 + \partial/\partial y(t_iU_i+t_jU_j+t_kU_k)^2] \\ +(t_iU_i+t_jU_j+t_kU_k)f(x,y) \}dxdy.$$

Differentiating this expression with respect to U_i and using the definition of t_p, $p=i,j,k$, gives

$$\begin{aligned}\partial I_t/\partial U_i &= [(y_j-y_k)U_i+(y_k-y_i)U_j+(y_i-y_j)U_k](y_j-y_k)/4\Delta^2 \\ &+ [(x_k-x_j)U_i+(x_i-x_k)U_j+(x_k-x_j)U_k](x_k-x_j)/4\Delta^2 \\ &\iint_t\{f(x,y)[(x_jy_k-x_ky_j)+(y_j-y_k)x+(x_k-x_j)y]\}/2\Delta \, dxdy,\end{aligned}$$

where Δ is the area of the relevant triangle. If the elements are sufficiently small and f does not change too rapidly we may replace f(x,y) by f evaluated at the centroid of the element, then

$$\partial I_t/\partial U_i = \alpha_iU_i + \alpha_jU_j + \alpha_kU_k + \beta f,$$

where

$$\begin{aligned}\alpha_i &= [(y_j-y_k)^2 + (x_k-x_j)^2]/4\Delta, \qquad \beta=\Delta/3, \\ \alpha_j &= [(y_k-y_i)(y_j-y_k)+(x_i-x_k)(x_k-x_j)]/4\Delta, \\ \alpha_k &= [(y_i-y_j)(y_j-y_k)+(x_j-x_i)(x_k-x_j)]/4\Delta.\end{aligned}$$

As a simple example consider the solution of Laplace's equation over a square covered by triangular elements as shown in Figure 6.6.2. The contribution from triangle number 1 in Figure 6.6.2 is given by

$$\partial I_1/\partial U_0 = (h^2U_0 - h^2U_4)/2h^2.$$

Similarly for the other triangles

$$\partial I_2/\partial U_0 = (U_0-U_3)/2, \qquad \partial I_3/\partial U_0 = (2U_0-U_2-U_3)/2,$$

$$\partial I_4/\partial U_0 = (U_0-U_2)/2, \qquad \partial I_5/\partial U_0 = (U_0-U_1)/2,$$

$$\partial I_6/\partial U_0 = (2U_0-U_1-U_4)/2,$$

and combining these terms gives the equation

$$U_1 + U_2 + U_3 + U_4 - 4U_0 = 0$$

which is the normal five point difference approximation, as obtained in Section 6.2, at the centre of the square with mesh spacing half the length of one side.

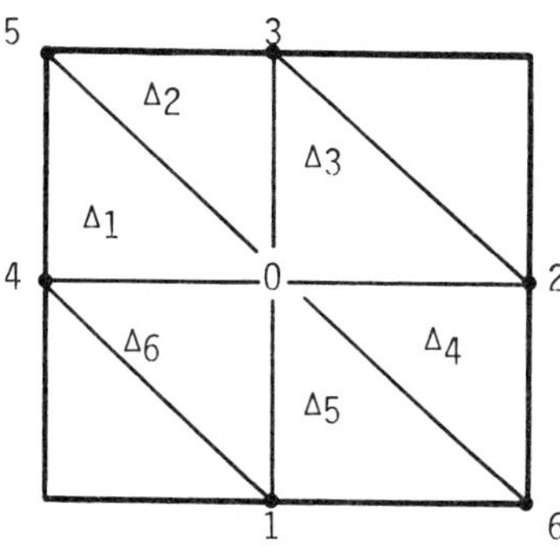

Figure 6.6.2 A simple triangularisation of a square.

The derivation of such equations using the finite element method is a little more complicated than that used in previous sections but this method has the advantage that the size of the elements need not be uniform, in particular they can be selected to take into account the shape of the boundary. Furthermore, any boundary conditions are satisfied automatically by the choice of basis functions. Against this must be weighed the problem of deriving the contribution to the functional from each element and also the lack of a simple error estimate. In general it may be necessary to perform any integration over the elements numerically. However, by selecting suitable trial functions it is possible to get very accurate results.

The finite element method can also be applied to initial value problems involving parabolic partial differential equations, however, the coefficients which are now to be determined will be functions of t. For example, the functional

$$J(u) = \iint_\Gamma [{}^1/_2(u_x)^2 + {}^1/_2(u_y)^2 + u\, u_t]\, dxdy$$

has a stationary value when $u_t = u_{xx} + u_{yy}$ throughout Γ and satisfies suitable conditions on the boundary of Γ. In general if the differential equation does not satisfy a known variational principle then we attempt to minimise the error corresponding to equation (6.6.4).

Example 6.6.3 Consider the solution of the non-linear parabolic partial differential equation $u_t = u_{xx} + F(u)$, $0 \leq x \leq 1$, $t > 0$. Let $w(x,t)$ be a trial solution then we attempt to minimise the expression

$$\int_0^1 f_i(x)[w_t - w_{xx} - F(w)]dx, \quad i=1,2,\ldots,N,$$

where N is the number of parameters at our disposal and $f_i(x)$ are suitable weight functions. If w is a linear combination of basis functions, i.e.

$$w(x,t) = \sum_{j=1..N} c_j(t)\, \phi_j(x)$$

then

$$\sum_{j=1..N} \int_0^1 [\phi_i(x) c_j' \phi_j(x) + c_j \phi_j' \phi_i' - \phi_i F]dx = 0, \quad i=1..N,$$

gives a Galerkin method for determining the coefficients $c_j(t)$, $j=1,..,N$. In matrix notation this system of equations can be written as

$$\mathbf{A}\mathbf{c}' = -\mathbf{B}\mathbf{c} + F(c),$$

where $c=[c_1(t),c_2(t),...,c_N(t)]^T$ and the elements of A, B and F are given by

$$A_{ij} = \int_0^1 \phi_j(x)\ \phi_i(x)\ dx, \qquad B_{ij}(x) = \int_0^1 \phi'_j(x)\ \phi'_i(x)\ dx,$$

$$F_i = \int_0^1 \phi_i F(w)\ dx.$$

This system of ordinary differential equations can then be solved numerically to determine the coefficients c(t). Choosing the basis functions ϕ_j to be the cubic splines given in equation (3.5.4) reduces both A and B to tridiagonal matrices. Further details are given by Jain (1977), Mitchell and Griffiths (1977) and Zienkiewicz and Morgan (1983).

APPENDIX 1
The Algebraic Eigenvalue Problem

Let A be an n×n matrix, **x** a non-zero vector of dimension n and λ a scalar then a solution $(\mathbf{x},\lambda)$ of the equation

$$\mathbf{Ax} = \lambda\mathbf{x} \qquad (A.1.1)$$

is called an eigensolution of A, **x** is called an eigenvector and λ is the associated eigenvalue. Re-arranging equation (A.1.1) gives

$$(\mathbf{A} - \lambda\mathbf{I})\mathbf{x} = 0,$$

where I is the identity matrix of dimension n, and since **x** is non-zero we conclude that the matrix $\mathbf{A}-\lambda\mathbf{I}$ is singular, therefore, $\mathrm{Det}(\mathbf{A}-\lambda\mathbf{I})=0$. If this determinant were multiplied out we would obtain a polynomial equation of degree n in λ and so by the Fundamental Theorem of Algebra there exist exactly n values for λ. If these values are distinct then for each one there will be exactly one independent eigenvector. If they are not distinct the situation is more complicated. Notice that if x is an eigenvector associated with λ then any scalar multiple of x is also an eigenvector. To avoid confusion it is usual to normalize the vector so that $\|\mathbf{x}\|=1$. Full and complete details are given in many books on Linear Algebra. (For example, see Liebeck (1969) or Wilkinson (1965).)

The determination of even a single eigenvalue is by no means trivial and if the dimension of **A** exceeds 2 then reducing the problem (A.1.1) to solving a polynomial equation invites trouble. (See Wilkinson (1965).) However, it is possible to find the eigenvalues of largest and smallest modulus by the Power Method as was demonstrated in Section 1.4. This type of technique can be extended to finding other eigenvalues by looking at matrices of the form $\mathbf{A}-p\mathbf{I}$, in which case inverse iteration gives the eigenvalue of A which is closest to p. If more than one eigenvalue, or even all the eigenvalues, of A are required then they can be found by reducing the matrix **A** to a more compact form. Theoretically, there exists a sequence of similarity transformations which reduce **A** to a diagonal matrix if all the eigenvalues are distinct or a Jordan canonical form if not. In practice a more stable procedure is to reduce **A** to an upper Hessenberg form, (upper triangular plus one diagonal immediately below the main diagonal), using similarity transformations and then find the eigenvalues by the QR algorithm or one of its many variants. Full details of this method and a comprehensive error anaylsis are given by Wilkinson (1965).

Certain matrices, which appear quite frequently, have a very simple set of eigenvalues. For example, let A be the n dimensional tridiagonal matrix

$$\mathbf{A} = \begin{bmatrix} b & c & & & & \\ a & b & c & & & \\ & a & b & c & & \\ & & \cdot & \cdot & \cdot & \\ & & & \cdot & \cdot & \cdot \\ & & & a & b & c \\ & & & & a & b \end{bmatrix} . \qquad (A.1.2)$$

The eigenvalues of A are given by

$$\lambda_j = b + 2\{\surd(ac)\} \cos{}^{(j\pi)}/_{(n+1)}, \quad j=1,\ldots,n$$

and the corresponding eigenvector x_j has as its kth component $\{^c/_b\}^{k/2} \sin\{^{(jk\pi)}/_{(n+1)}\}$.

Unfortunately, most matrices do not reveal their eigenvalues so easily, but we can estimate their location by using the following result.

Theorem A.1.1 (Gershgorin's Theorem) Every eigenvalue of a matrix **A** lies in at least one of the discs:

$$D_i = \{ \lambda \varepsilon \mathbf{C} : |\lambda - a_{ii}| \leqslant \sum_{\substack{j=1..n \\ i\neq j}} |a_{ij}| \}$$

Proof. Let λ be any eigenvalue of A for which the corresponding non-zero eigenvector is **x**. Suppose that the kth component of **x** has the largest modulus then we may normalize x by dividing **x** by x_k to give

$$\mathbf{x} = \lfloor x_1,\ldots,x_{k-1},1,x_{k+1},\ldots,x_n\rfloor^T$$

where

$$|x_j| \leqslant 1, \quad j=1,\ldots,n.$$

We now examine the equation $\mathbf{Ax} = \lambda\mathbf{x}$ in its component form, i.e.

$$\sum_{j=1..n} a_{kj}\, x_j = \lambda x_k = \lambda.$$

Therefore,

$$|\lambda-a_{kk}x_k| = |\lambda-a_{kk}| \leqslant \sum_{j\neq k} |a_{kj}x_j| \leqslant \sum_{j\neq k} |a_{kj}|\; |x_j| \leqslant \sum_{j\neq k} |a_{kj}|,$$

which implies that λ lies in the disc D_k.

Example A.1.1 Let A be the matrix

$$\begin{bmatrix} 1 & 0.25 & 0 \\ 0.25 & 2 & -0.25 \\ 0 & 0.25 & 3 \end{bmatrix}$$

then

$$D_1 = \{ \lambda \in \mathbf{C} : |\lambda - 1| \leqslant 0.25 \},$$
$$D_2 = \{ \lambda \in \mathbf{C} : |\lambda - 2| \leqslant 0.5 \},$$
$$D_3 = \{ \lambda \in \mathbf{C} : |\lambda - 3| \leqslant 0.25 \},$$

therefore, the eigenvalues of A lie within the shaded area shown in Figure A.1.1.

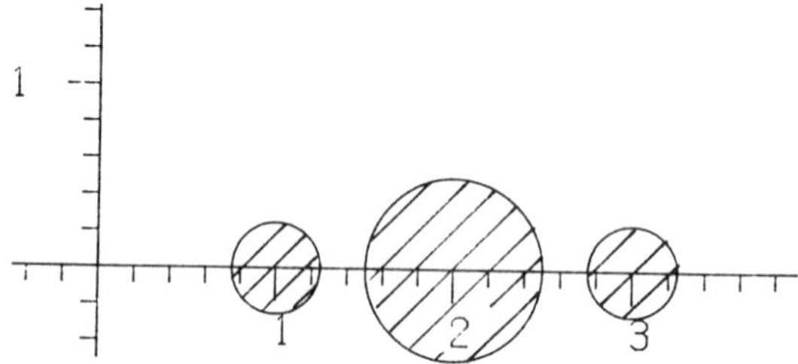

Figure A.1.1 **The discs D_i, i=1,2,3, for Example A.1.1.**

Example A.2.2 If a matrix A is strictly diagonally dominant then it is non-singular.

Solution. Since A is strictly diagonally dominant we have that

$$|a_{ii}| > \sum_{i \neq j} |a_{ij}|, \qquad i=1,2,\ldots,n$$

Therefore,

$$|\lambda - a_{ii}| < |a_{ii}|, \; i=1,\ldots,n$$

and since all the eigenvalues of A are bounded away from zero we conclude that A is non-singular.

APPENDIX 2
The Classification of Partial Differential Equations

Let $u(x,y)$ be a continuous function which maps R^2 into R and u_x and u_y be its first partial derivatives with respect to the variables x and y respectively. Any relationship of the form

$$\Psi(x,y,u,u_x,u_y) = 0, \qquad (A.2.1)$$

is called a first order partial differential equation. Similarly, let u_{xx}, u_{xy} and u_{yy} be the second partial derivatives of u,in an obvious notation then

$$\Psi(x,y,u,u_x,u_y,u_{xx},u_{xy},u_{yy}) = 0, \qquad (A.2.2)$$

is called a second order partial differential equation. (It is assumed that the function is sufficiently differentiable to ensure that $u_{xy}=u_{yx}$.) The equation (A.2.2) is very general and so for the sake of clarity we shall consider only equations of the form

$$au_{xx} + bu_{xy} + cu_{yy} + du_x + eu_y + fu + r = 0. \qquad (A.2.3)$$

If the coefficients a, b, c, d, e and f are only functions of x and y then this equation is said to be linear. When the coefficients of the second derivatives are functions of x,y,u and the first partial derivatives whilst the coefficients of u_x and u_y are functions of x,y and u, and f and r are functions of x and y alone the equation (A.2.3) is said to be quasi-linear.

The general solution of an n-th order partial differential equation will involve n arbitrary functions, if these are specified then we obtain a particular solution. This means that if we are to obtain a particular solution we will need to specify additional restrictions in addition to the differential equation. Such restrictions fall into two main categories. The first type concerns the solution of the differential equation in a closed bounded region where the solution or its derivative is specified at every point of the boundary. Such problems are called **Boundary Value Problems.** The second type of problem concerns the solution of the differential equation on a possibly semi-infinite region subject to some form of initial condition on part of the boundary and the solution then proceeds by marching through the rest of the region subject to any other boundary conditions. These problems are called **Initial Boundary Value Problems**, Initial Value or Marching problems.

Having briefly classified the problem we now consider the classification of the differential equation itself. Let us suppose that u, u_x and u_y are known along a curve Γ then, by expanding in a double Taylor series about (x,y), the solution at (x+h,y+k) can be written as

$$u(x+h,y+k)=u(x,y) + hu_x + ku_y + {}^{h^2}/_2 u_{xx} + hku_{xy} + {}^{k^2}/_2 u_{yy} + \dots$$

If u(x,y), $u_x(x,y)$ and $u_y(x,y)$ are known then we can use the above expression to estimate u(x+h,y+k). To do this we examine the derivatives of u, u_x and u_y along the direction γ as shown in Figure A.2.1.

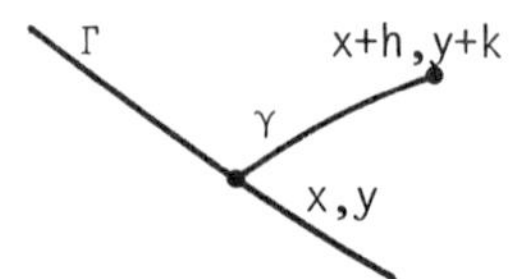

Figure A.2.1 **The curves Γ and γ.**

Let D denote differentiation along the direction γ then

$$D(u_x) = u_{xx}\,dx + u_{xy}\,dy, \qquad (A.2.4)$$
$$D(u_y) = u_{xy}\,dx + u_{yy}\,dy. \qquad (A.2.5)$$

We collect together the equations (A.2.3), (A.2.4) and (A.2.5) to give

$$\begin{bmatrix} a & b & c \\ dx & dy & 0 \\ 0 & dx & dy \end{bmatrix} \begin{bmatrix} u_{xx} \\ u_{xy} \\ u_{yy} \end{bmatrix} = \begin{bmatrix} -du_x-eu_y-fu-r \\ D(u_x) \\ D(u_y) \end{bmatrix}$$

which determine u_{xx}, u_{xy} and u_{yy} provided these equations are soluble. Whether this is the case, or not, depends upon the determinant of the matrix, if it is non-zero then there will be exactly one solution. The determinant will vanish if

$$a(dy)^2 - bdydx + c(dx)^2 = 0,$$

i.e. when

$$a({}^{dy}/_{dx})^2 - b\,{}^{dy}/_{dx} + c = 0. \qquad (A.2.6)$$

There are three distinct cases which can occur:-

(i) **Two real roots** in which case equation (A.2.6) gives two possible directions

$$^{dy}/_{dx} = \left(-b \pm \sqrt{(b^2-4ac)}\right)/_{2a}$$

for the direction γ. If this is the case then the partial differential equation is said to be **hyperbolic.** The two directions given by (A.2.6) play a fundamental role in the solution of partial differential equations of this type and are called **characteristics.** Any discontinuity which exists will propagate along such lines.

(ii) **Two coincident roots** in which case the equation is said to be **parabolic.**

(iii) **Two complex roots** in which case there are no directions γ along which the solution cannot be expanded and the differential equation is said to be **elliptic.**

Example A.2.1

(i) The Wave Equation $u_{xx}=c^2u_{yy}$ is hyperbolic.

(ii) The Heat or Diffusion Equation $u_y=\sigma u_{xx}$ is parabolic.

(iii) Laplaces equation $u_{xx}+u_{yy}=0$ is elliptic.

Example A.2.2 Show that the equation

$$xu_{xx} - u_{yy} + {}^1/_2\, u_x = 0$$

is hyperbolic if $x<0$, parabolic when $x=0$ and elliptic if $x>0$.

Solution. The equation which governs the characteristics for this problem is given by (A.2.6) which becomes

$$x\left({}^{dy}/_{dx}\right)^2 - 1 = 0,$$

or ${}^{dy}/_{dx}=\pm{}^1/_{\sqrt{x}}$ which gives real solutions for $x>0$ and so the differential equation is hyperbolic in this region. For $x<0$ there are no real characteristics and so the equation is elliptic.

Returning to equation (A.2.6) which determines the classification of partial differential equations of the type (A.2.3) we can use the characteristic directions to reduce the equation to canonical forms. Integrating (A.2.6) will produce two families, possibly coincident, of curves which involve two constants of integration ξ and η which determine to which member of each family each point belongs. If we change variables to $\xi(x,y)$ and $\eta(x,y)$ we obtain the natural or characteristic co-ordinate system for the differential equation.

Example A.2.3 Determine the characteristics for the equation

$$xu_{xx} - u_{yy} + {}^1/_2 u_x = 0$$

and reduce it to its canonical form.

Solution. The characteristic curves are given from equation (A.2.6) and satisfy ${}^{dy}/_{dx}={}^1/_{\sqrt{x}}$ and ${}^{dy}/_{dx}=-{}^1/_{\sqrt{x}}$. The first family of characteristic curves is given by

$$y = 2\sqrt{x} + \xi$$

and the second by

$$y = -2\sqrt{x} + \eta.$$

Using the characteristic variables $\xi=y-2\sqrt{x}$ and $\eta=y+2\sqrt{x}$ the given differential equation becomes

$$\partial^2 u/_{\partial\xi\partial\eta} = (u_{yy} - xu_x - {}^1/_2 u_x) = 0.$$

This is the canonical form for this equation and has as its solution $u(\xi,\eta)=F(\xi)+G(\eta)$, where F and G are arbitrary functions. The solution of the original differential equation is, therefore, given by

$$u(x,y) = F(y-2\sqrt{x}) + G(y+2\sqrt{x}).$$

The change to characteristic variables shown in Example A.2.3 demonstrates that the only second derivative which appears is $\partial^2 u/_{\partial\xi\partial\eta}$. In general, if the equation (A.2.3) is linear then changing to a characteristic co-ordinates system will produce the canonical partial differential equation

$$\partial^2 u/_{\partial\xi\partial\eta} = \Phi(\xi,\eta,u,u_\xi,u_\eta). \qquad \text{(A.2.7)}$$

If the partial differential equation is parabolic then the two families of characteristics are one and the same and a change to this co-ordinate system will reduce the differential equation to the canonical form

$$\partial^2 u/\partial\eta^2 = \Phi(\xi, \eta, u, u_\xi, u_\eta). \qquad (A.2.8)$$

Finally, if the differential equation is elliptic the mixed derivative can be eliminated by changing to the characteristic co-ordinate system in which case the equation becomes

$$\partial^2 u/\partial\xi^2 + \partial^2 u/\partial\eta^2 = \Phi(\xi, \eta, u, u_\xi, u_\eta). \qquad (A.2.9)$$

For further details on the role that characteristics play see for example Weinberger (1965) or Ames (1977).

Solutions for Exercises in Chapter 1

1.1.1 (i) $y_n = Ar^n$, A arbitrary.

(ii) $r \neq 1$, $y_n = Ar^n + {}^n/_{(1-r)} - {}^1/_{(1-r)^2}$,

$r=1$, $y_n = 0.5(n^2 - n) + A$, A arbitrary.

(iii) $y_n = (-1)^n A/_{(n-1)} + 0.5n$, $n>1$, A arbitrary.

1.1.2 $\alpha=0$: $y_n=n+1$. $\alpha\neq 0$: $y_n = {}^1/_\alpha((\alpha+1)^{n+1} - 1)$.

$-2<\alpha<0$: absolutely and relatively stable,

$\alpha>0$ or $\alpha<-2$: absolutely unstable but relatively stable.

1.2.1 (i) $y_n = A(-3)^n + B2^n$, A and B arbitrary.

(ii) $y_n = (-1)^n (A + Bn)$, A and B arbitrary.

1.2.2 (i) $y_n = A(-3)^n + B2^n + {}^5/_4 - n^2$.

(ii) $y_n = A({}^{(1+\sqrt{5})}/_2)^n + B({}^{(1-\sqrt{5})}/_2)^n$.

1.2.3 As Example 1.2.2 with the arbitrary constants given by

(i) $A = {}^{-9}/_{20}$, $B = {}^{-16}/_{20}$.

(ii) $A = {}^{-23}/_{156}$, $B = {}^{-45}/_{130}$.

(iii) $A = 1.25\ (79+2^{10})/_{(3^{10}-2^{10})}$, $B=-1.25-A$.

1.2.4 (i) $y_n = 0.5[(0.1)^n - (-0.1)^n]$.

(ii) $y_n = A[(0.1)^n - (-0.1)^n]$, A arbitrary.

1.2.5 Example 1.2.3 is unstable, Example 1.2.4 is stable.

1.3.1 Both converge linearly to 2.

1.3.2 There are 3 real solutions 1, 0.741270911, -1.44685725

1.3.4 (i) 1.93456 radians.

(ii) 1.14619322, -1.84140566.

1.3.5 The double root is at x=1.463055513 when k=4.319136566.

1.5.8 Trapezium rule requires approximately 3700 intervals.
Simpson's rule requires approximately 64 intervals.
Romberg extrapolation gives the value 1.462652 to 6 d.p.

Literature

Ames, W.F. (1977) Numerical Methods for Partial Differential Equations. 2nd Edition. Nelson.

Bender, C.M. and Orszag, S.A. (1978) Advanced methods for scientists and engineers. McGraw Hill.

Bers, L., John, F and Schechter, M. (1964) Partial Differential Equations. Interscience, John Wiley.

Brezinski, C. (1978) Algorithmes d'accéleration de la convergence. Etudes numérique, Technip, Paris.

Burden, R.L., Faires, J.D. and Reynolds, A.C. (1981). Numerical Analysis. 2nd Edition. Prindle, Weber and Schmidt. (Wadsworth).

Butcher, J. (1964) Implicit Runge-Kutta processes. Math.,Comp., vol **18**, 50-64.

Cash, J.R. (1984) Two new finite difference schemes for parabolic equations. SIAM J. Numer. Anal. Vol 21, No.3.

Cheney, W. and Kincaid D. (1980) Numerical Mathematics and Computation. Brooke/Cole Publishing Co., (Wadsworth).

Coddington, E.A. and Levinson, N. (1955) Theory of Ordinary Differential Equations. McGraw Hill.

Collatz, L. (1960) The numerical treatment of differential Equations. Berlin, Springer.

Daniel, J.W., and Moore, R.E. (1969) Computation and Theory in Ordinary Differential Equations. W.H.Freeman and Company. San Francisco.

Davis, P.J. and Rabinowitz, P. (1975) Methods of Numerical Integration. Academic Press, New York.

Davis, H.T. (1960) Introduction to Nonlinear Differential and Integral Equations. Dover.

Deuflhard, P. (1980) Recent advances in multiple shooting techniques. In **Gladwell, I. and Sayers, D.K.** (1980) Computation Techniques for Ordinary Differential Equations. Academic Press.

Elsgolc, L.E. (1963) Calculus of Variations. Pergamon.

Erisman, A and Reid, J. (1981) Direct Methods for Sparse Matrices. Oxford Unversity Press, Oxford.

Fox, L. (1962) The Numerical Solution of Ordinary and Partial Differential Equations. Oxford, Pergamon Press, 106-111.

Fox, L. (1964) Introduction to Numerical Linear Algebra. Oxford.

Fox, L. and Mayers, D.F. (1968) Computing methods for Scientists and Engineers. Oxford University Press.

Fox, L. and Parker, I.B. (1972) Chebyshev Polynomials in Numerical Analysis. Oxford Mathematical Handbooks.

Fox, L. (1977) Finite Difference methods for Elliptic Boundary Value Problems. In **Jacobs, D.A.H.** The State of the Art in Numerical Analysis, Academic Press.

Gahkov, F.D. (1966) Boundary Value Problems. (Translated by I.N. Sneddon). Pergamon.

Garabedian, P.R. (1964) Partial Differential Equations. John Wiley and Sons, Inc., New York.

Gear, C.W. (1971) Numerical Initial Value Problems in Ordinary Differential Equations. Prentice Hall, N.J.

George, J.A. (1973) Nested dissection of regular finite element meshes. SIAM. J.Numer.Anal. **10**, 345-363.

Gerald, C.F. (1977) Applied Numerical Analysis. 2nd Edition. Addison Wesley.

Gladwell, I. and Sayers, D.K. (1980) Computational Techniques for Ordinary Differential Equations. Academic Press.

Gourlay, A.R. and Morris, J.Ll. (1981) Linear combinations of generalised Crank-Nicolson schemes. IMA J. Numer. Anal. **1**, pp347-357.

Hagan, P.S. (1981) The Instability of Nonmonotonic Wave Solutions of Parabolic Equations. Studies in Applied Maths. **64**,57-88.

Hall, G. and Watt, J.M. (1976) Modern Numerical Methods for Ordinary Differential Equations. Clarendon Press, Oxford.

Henrici, P. (1968) Discrete variable methods in Ordinary Differential Equations. John Wiley.

Hubbard, B. (1966) Remarks on the order of the convergence in the discrete Dirichlet problem. In "Numerical solution of partial differential equations." Ed. J.H.Bramble, Academic Press, New York and London.

Hosking, R.J, Joyce, D.C. and Turner, J.C. (1978) First Steps in Numerical Analysis. Hodder and Stoughton.

Jacobs, D.A.H. (1977) The State of the Art in Numerical Analysis. Academic Press, London.

Jain, M.K. (1979) Numerical Solution of Differential Equations. Wiley Eastern Ltd. New Delhi.

Jones, D.S. and Sleeman, B.D. (1983) Differential Equations and Mathematical Biology. Allen and Unwin.

Jordan, D.W. and Smith, P. (1977) Non-linear Ordinary Differential Equations. Oxford Applied Mathematics and Computing Series.

Keller, H. (1968) Numerical methods for two point boundary value problems. Waltham, Balisdell.

Lambert, J. (1973) Computing methods for Ordinary Differential Equations. Wiley, N.Y.

Lawson, J.D. and Morris J.Ll. (1978) The extrapolation of first order methods for parabolic partial differential equations, I. SIAM J.Numer.Anal, Vol 15, pp1212-1224.

Lax, P.D. (1954) Weak solutions of non-linear hyperbolic equations and their numerical computation. Comm.Pure Appl. Math., **7**, 157-193.

Lefschetz, S. (1968) Differential Equations: Geometric Theory. 2nd Edition. John Wiley.

Liebeck, H. (1969) Algebra for Scientists and Engineers. John Wiley and Sons, London.

Lyusternik, L.A. (1947) A note for the numerical solution of boundary value problems for the Laplace equation and for the calculation of eigenvalues by the method of nets. Trudy Inst. Math. Academy of Sciences of the USSR, **20**, 49-64.(Russian).

Marchuk, G.I. and Shaidurov, V.V. (1983) Difference Methods and Their Extrapolations. Springer-Verlag, N.Y.

Meis, T. and Marcowitz, U. (1981) Numerical Solution of Partial Differential Equations. Applied Mathematical Sciences 32. Springer-Verlag. New York.

Mitchell, A.R. and Griffiths, D.F. (1977) The Finite Element Method in Partial Differential Equations. (Second Edition). J.Wiley.

Murray, J.D. (1976) On a model of concentration waves in the Belousov-Zhabotinskii reaction. J.theor. Biol. **56**, 329-353.

Noble, B. (1966) Numerical Methods 2: Differences, Integration and Differential Equations. Oliver and Boyd.

Nordsieck, A. (1962) "On numerical integration of ordinary differential equations." Math.Comp. 16. 22-49.

Ortega, J.M. and Poole, W.G. Jnr. (1981) An Introduction to Numerical Methods for Differential Equations. Pitman Publishing Inc.

Osborne, M.R. (1969) On shooting methods for boundary value problems. J.Math.Anal.Appl. **27**, 417-433.

Parker, I.B. and Crank, J. (1964) Persistent discretization error in partial differential equations of parabolic type. Computer J., **7**, 163-167.

Philips, G.M. and Taylor, P.J. (1973) Theory and Application of Numerical Analysis. Academic Press.

Pizer, S.M. (with Wallace, V.L.) (1983) To Compute Numerically. Little, Brown Computer System Series, Boston.

Quinney, D.A. (1979) On Computing Travelling Wave Solutions in a Model for the Belousov-Zhabotinskii reaction. J. Inst.Math.Applics. **23**, 193-201.

Reid, J.K. (1977) Sparse matrices. In **Jacobs, D.A.H.** The State of the Art in Numerical Analysis, Academic Press.

Reiss, E.L, Callegari, A.J. and Ahluwalia, D.S. (1976) Ordinary Differential Equations with Applications. Holt, Rinehart and Winston.

Richardson L.F. (1927) The deferred approach to the limit. Trans. Phil. Roy. Soc. 226 pp261-299.

Richtmyer, R.D. and Morton, K.W. (1967) Difference Methods for Initial Value Problems. John Wiley.

Rosinger, E.E. (1980) Stability and Convergence of Non-linear Difference Schemes are Equivalent. J.Inst.Math.Applics. **26**, 143-149.

Sawyer, W.W. (1977) A First look at Numerical Functional Analysis. Oxford Applied Mathematics and Computing Series.

Shampine, L. and Allen, R. (1973) Numerical Computing: An Introduction. Saunders, Philadelphia.

Shampine, L. and Gear, C.W. (1979) A User's view of solving Stiff Ordinary Differential Equations. SIAM Review 21, 1-17.

Smith, G.D. (1978) Numerical Solution of Partial Differential Equations: Finite Difference Methods. Oxford Applied Mathematics and Computing Series, OUP.

Stephenson, G. (1970) An Introduction to Partial Differential Equations. 2nd Edition. Longman.

Stoer, J and Burlisch, R. (1980) Introduction to Numerical Analysis. Springer-Verlag.

Strang, G and Fix, G.T. (1973) An analysis of the Finite Element Method. Prentice Hall, Englewood Cliffs, N.J.

Tierney, J.A. (1979) Differential Equations. Allyn and Bacon.

Varga, R.S. (1962) Matrix Iterative Analysis. Prentice Hall.

Weinberger, H.F. (1965) A First Course in Partial Differential Equations. Xerox College Publishing, Toronto.

Whiteman, J.R. (Editor) (1972) The Mathematics of Finite Elements and Applications. Academic Press.

Widlund, D.E. (1967) "A note on unconditionally stable linear multistep methods." BIT, **7**, pp65-70.

Williams, P.W. (1973) Numerical Computation. Nelson.

Wilkinson, J.H. (1965) The Algebraic Eigenvalue Problem. OUP.

Young, D. (1971) Iterative Solution of Large Linear Systems. Academic Press, New York.

Zienkiewicz, O.C. and Morgan, K. (1983) Finite Elements and Approximation. Wiley International.

Index

Index